BROWN BEARS IN ALASKA'S NATIONAL PARKS

BROWN BEARS

IN ALASKA'S NATIONAL PARKS

CONSERVATION OF A WILDERNESS ICON

Edited by

GRANT V. HILDERBRAND, KYLE JOLY, DAVID D. GUSTINE, AND NINA CHAMBERS

University of Alaska Press | FAIRBANKS

Copyright © 2025

Published by University of Alaska Press
An imprint of University Press of Colorado
1580 North Logan Street, Suite 660
PMB 39883
Denver, Colorado 80203-1942

The University Press of Colorado is a proud
member of Association of University Presses.

The University Press of Colorado is a cooperative publishing enter-
prise supported, in part, by Adams State University, Colorado School
of Mines, Colorado State University, Fort Lewis College, Metropol-
itan State University of Denver, University of Alaska Fairbanks,
University of Colorado, University of Denver, University of Northern
Colorado, University of Wyoming, Utah State University, and Western Colorado University.

∞ This paper meets the requirements of the ANSI/NISO Z39.48-1992
(Permanence of Paper).

ISBN: 978-1-64642-709-3 (hardcover)
ISBN: 978-1-64642-710-9 (paperback)
ISBN: 978-1-64642-711-6 (ebook)
https://doi.org/10.5876/9781646427116

Frontispiece: NPS/Mary Lewandowski
Cover photo: Katrina Marie Hill

Library of Congress Cataloging-in-Publication Data

Names: Hilderbrand, G. V. (Grant Vaughan), 1971– editor
Title: Brown bears in Alaska's national parks : conservation of a
    wilderness icon / edited by Grant V. Hilderbrand, Kyle Joly, David D.
    Gustine, and Nina Chambers.
Description: Fairbanks : University of Alaska Press, [2025] | Includes
    bibliographical references and index.
Identifiers: LCCN 2024058226 (print) | LCCN 2024058227 (ebook) |
    ISBN 9781646427093 hardcover | ISBN 9781646427109 paperback |
    ISBN 9781646427116 ebook
Subjects: LCSH: Kodiak bear—Alaska—History | Brown bear—
    Alaska—History | National parks and reserves—Alaska—
    History | Wildlife management—Alaska. | National parks and
    reserves—Management—History
Classification: LCC QL737.C27 B7566 2025 (print) | LCC QL737.C27
    (ebook) | DDC 599.78409798—dc23/eng/20250221
LC record available at https://lccn.loc.gov/2024058226
LC ebook record available at https://lccn.loc.gov/2024058227

▶ Alaska brown bear and salmon, an iconic combination. (Mark Johnson)

CONTENTS

Brown bears are adaptable, live in a variety of landscapes, and take advantage of available food sources, like clams in the intertidal zone. (NPS/Jim Pfeiffenberger)

FOREWORD

Brown Bears in Alaska's National Parks provides a vivid description of the importance of brown bears to Alaskans and to Alaska's natural environment. The book helps us better understand this iconic species and what is required for them to survive and thrive, as well as describing the ways they are being challenged by development and climate.

I have been fortunate to see brown bears in a variety of places in Alaska, from Pack Creek in the southeast to McNeil River in the south-central portion of the state, as well as in Alaska's wonderful national parks like Katmai, Lake Clark, and Glacier Bay. Sharing these experiences with fellow Alaskans as well as with tourists from all over the world has illustrated to me the extraordinary and unique value of these animals and the importance of habitat protection efforts and science-informed management to assure they can continue to thrive. Both federal and state agencies that make these management decisions rely on the information scientists provide. The public has an interest is assuring that there is adequate funding for the research and that unbiased information is used effectively.

The authors have spent many decades observing, monitoring, and researching brown bears in a variety of national parks, giving them a deep understanding of brown bear diets and habits specific to different regions in Alaska, as well as the ability to compare them. This book describes the complex relationship between people and brown bears, including the traditional and spiritual connections Alaska Native cultures embrace.

Brown bears are highly intelligent and adaptive animals, as demonstrated by the remarkable variability and flexibility within the species across a wide variety of landscapes. Even within a specific location, they can change their behaviors to cope with changing conditions. However, there are limits to their resilience; if we want to assure their continued presence, we need to understand those limits and how to avoid pushing bears beyond them. The science shared in this book demonstrates the importance of efforts to continue monitoring the species so that land management practices can be altered appropriately, based on that science.

It also reminds us as members of the public how important it is that our voices are heard advocating on behalf of the wildlife in our national parks.

FRAN ULMER

Lt. Governor of Alaska, 1994–2002
Chancellor, University of Alaska
Anchorage, 2007–2010
Chair, United States Arctic Research
Commission, 2011–2020
Trustee, National Parks Conservation
Association, 2004–present
Chair, Global Board of the Nature
Conservancy, 2019–present

▶ Brown bears are highly intelligent and much of their knowledge is transmitted from mothers to cubs. (NPS/Lian Law)

Alaska is a large, rugged territory for tracking bears. Pictured here are the Neacola Mountains, Aleutian Range, in Lake Clark National Park and Preserve. The National Park Service uses telemetry tracking for bears across the state. (NPS/Buck Mangipane)

PREFACE

National parks are home to many of Earth's most striking landscapes and treasured natural resources. The character that makes each park significant existed long before governments established boundaries delineating them. These areas, the ecosystems that comprise them, and the wonder they have instilled in humans predate recorded history.

The four of us are incredibly fortunate to work in Alaska's national parks. We view ourselves as stewards of resources that belong to all of us. Parks are something we all share but none of us possess. Working in parks, one feels insignificant in both time and space as the vistas are vast and time is measured in epochs. It feels both humbling and comforting to be part of something so much bigger than ourselves. We have yet to have a day in the field when we didn't see something new and amazing. One purpose of this book is to share some of our moments of discovery and amazement with you.

Brown bears are, perhaps, the most iconic North American symbol of wild things and wild places. They have inspired folklore, fear, wonder, and controversy. In some ways, the battle of nature versus progress has been waged around brown bears for centuries. In Alaska, we are blessed to have healthy populations of brown bears that span the state's wildly diverse, expansive, and largely intact natural ecosystems. Wildlife generally, and bears specifically, is why many national parklands were created in Alaska.

In this book, we discuss aspects of the natural history, ecology, and behavior of brown bears. In addition, we outline bear safety in a somewhat cursory manner because other, more definitive books and resources exist on this topic. Our second purpose for writing this book is to consolidate and describe the research we and our colleagues have conducted in Alaska parks in recent decades. As scientists, our currency is scientific publications that report our findings from very specific and nuanced studies. Here, we synthesize the knowledge gained through numerous national park studies holistically. We did not attempt to synthesize all current research on brown bears and note that many other people and agencies are also conducting high-quality research on them. As fundamental as discovery is to science, of equal importance are the new questions that arise as we learn. We also look to chart a path toward future needed work. This is of particular importance given how dynamic our world is due to the rapid and immediate effects of climate change and development.

A third purpose for this book is to use bears as a model for how we, as scientists, identify appropriate research questions based on the challenges and opportunities that face national park managers.

These questions vary from park to park and across time. Succinctly, we know far less about most bear populations than one would likely suspect. Bears are challenging and expensive to study. Thus, we have to be strategic as we refine our research questions, develop and implement our projects, analyze and publish our results, and share our findings with those who decide how parks and the resources they were created to protect are managed. Further, we describe what we have learned with visitors and those who share our love of both the lands and the animals we strive to conserve.

Our final goal is, perhaps, the most important. It is simply to share our knowledge and admiration with you, so you join us in becoming stewards of this magnificent species. Understanding is a key to conservation. As brown bears are at the top of the food chain and require large, relatively undeveloped ranges, they will require stewards, like you, for them to persist side by side with us.

We know we are privileged to have made a career out of, what is for us, an innate passion. As much as anything, being a scientist is about the quest for knowledge, a thirst for discovering the unknown. If you picked up this book to thumb through it, read it, or give it as a gift, we know you or someone you love shares our passion. The natural world needs advocates, today more than ever. Thank you. We are honored to share a bit of our journey of discovery with you.

ACKNOWLEDGMENTS

The development of a book like this takes a huge team effort. We thank the scores of collaborators from all the individual bear projects who made the work possible. Special thanks to all the pilots who made our projects possible but especially Troy Cambier, the helicopter pilot who safely flew countless hours for many of the projects and provided a wealth of knowledge, experience, and humor to all of them. In science, it is often said "we stand on the shoulders of giants." Our initial knowledge was gained from and current work builds upon the foundation developed by many pioneering brown bear biologists; we are deeply indebted to them. We thank Marty Byrne and Angie Southwould for developing the maps. We thank all of our supervisors who saw the value of our involvement with this project, especially Sarah Creachbaum, Jeff Rasic, and Jim Lawler. Staff at the University of Alaska Press greatly improved the draft manuscript into this final product. We also thank two anonymous bear experts for reviewing and improving our initial draft.

BROWN BEARS IN ALASKA'S NATIONAL PARKS

◂ Catching a quick nap after fishing. (NPS/Kelsey Griffin)

Humans have a complicated relationship with brown bears at various times, characterized by respect, fear, admiration, disdain, and worship. (NPS/Ken Conger)

1 STEWARDSHIP

PRESERVING BEARS FOR FUTURE GENERATIONS

Grant V. Hilderbrand, Andee Sears, David Payer, and Nina Chambers

As long as humans and brown (or grizzly) bears have overlapped in space and time, their coexistence has been complex. It is a relationship characterized at various times by respect, fear, admiration, disdain, and even worship. Brown bears and humans shared the landscape long before the existence of modern countries or borders or laws. They lived and died with the land. Together they subsisted and survived solely from what their ecosystems provided. In many ways, bears and humans were the 2 most dominant species in western North America. While humans and bears had singular skirmishes, both lived broadly in a tense but tolerant truce.

"The mountains have always been here, and in them, the bears."
—RICK BASS

"If all the beasts were gone, men would die of a great loneliness of spirit."
—CHIEF SEATTLE

BEARS AND WESTERN CONCEPTS OF CONSERVATION

Following the arrival of European settlers, modern civilization spread westward with a worldview that nature and the land were to be tamed. This perspective and many of the associated practices tied to farming, logging, damming waterways, road building, and bounty hunting of predators greatly diminished

https://doi.org/10.5876/9781646427116.c001

both the numbers and the range of brown bears in North America. In time, however, recognition of the importance of protecting wild landscapes, species conservation, clean air, and clean water led to 2 conservation eras. The first was in the early 1900s, during which the National Park Service (NPS) was established (1916) along with national forests, parks, and refuges. The second era, in the 1960s and 1970s, resulted in numerous legislative acts including the Wilderness Act (1964), National Environmental Protection Act (1967), Clean Air Act (1970), Clean Water Act (1972), and Endangered Species Act (ESA, 1973).

Our professional discipline of wildlife management followed on the heels of the first conservation era. In the late 1970s, the field of conservation biology emerged as a further maturation of wildlife management that recognizes the complexity of ecological interactions and the forces that shape them. Brown bears in the contiguous United States were largely restricted to national parks at the time of their listing as "threatened" under the ESA in 1975. They are, arguably, the single-most symbolic and controversial icon of wilderness of the modern conservation era. Their natural history requires large tracts of relatively undisturbed land, connectivity between patches of suitable habitat, abundant nutritional resources, and, because of their inherently low reproductive rate, protection from excessive human-caused mortality.

For these reasons, brown bears were often identified as "keystone" or "umbrella" species. A tenet of the practice of conservation biology, this is the idea that if brown bear populations are healthy, the ecosystem is likely healthy because the wild conditions they require also benefit other components of the ecosystem.

THE NATIONAL PARK SERVICE CONSERVATION MISSION

We, as an agency, have a central role in wildlife conservation. The mission provided in the Organic Act of 1916 that created the NPS specifically includes a call for us to conserve wildlife: "[to] conserve the scenery and the natural and historic objects and the wild life therein and to provide for the enjoyment of the same in such manner and by such means as will leave them unimpaired for the enjoyment of future generations."[1]

The NPS interprets and implements this conservation mission in a manner "to understand, maintain, restore, and protect the inherent integrity of the natural resources, processes, systems, and values of parks."[2] The term *park* is generally used to mean a unit of the National Park System, including national parks, preserves, and monuments. In these areas, the NPS strives "to maintain all the components and processes of naturally evolving park ecosystems,

▶ Bears are often considered indicator species; that is, if bear populations are healthy, it is likely that other components of the ecosystem are also healthy. (NPS/Matt Harrington)

ANILCA and Its Interpretation by Federal and State Agencies

While some parks in Alaska were established in the early twentieth century, most were established or enlarged in 1980 by the Alaska National Interest Lands Conservation Act (ANILCA).[3] ANILCA set aside more lands for conservation than had any previous conservation legislation—more than 100 million acres. It more than doubled the size of the National Park System, almost tripled the size of the National Wildlife Refuge System, and nearly quadrupled the size of the National Wilderness Preservation System. It also added 3.3 million acres to the National Forest System and designated 25 waterways as Wild or Scenic under the Wild and Scenic Rivers Act of 1968. ANILCA added 12 new parks, 6 of which are national preserves. It also expanded 3 existing parks: Denali (formerly Mount McKinley), Glacier Bay, and Katmai. Overall, the vast acreage ANILCA added to the federal conservation estate facilitates wildlife conservation on an unprecedented scale. Moreover, Alaska parks still contain many of the species that existed following the last ice age.

Like the NPS Organic Act, ANILCA prominently highlights the importance of wildlife conservation. The first sentence of the law reads: "In order to preserve for the benefit, use, education and inspiration of present and future generations certain lands and waters in the State of Alaska that contain nationally significant natural, scenic, historic, archaeological, geological, scientific, wilderness, cultural, recreational, and *wildlife* values, the units described in the following titles are hereby established" (emphasis added). ANILCA further outlines the "intent and purpose" to manage wildlife in accordance with "recognized scientific principles and the purposes for which each conservation system unit was established." Brown bears inhabit all parks in Alaska; for ten of the newly created parks, the purposes for which they were established expressly include protection of "habitat for, and populations of," brown or grizzly bears.

ANILCA also acknowledges the importance of wildlife harvest on Alaska parklands. One of ANILCA's fundamental tenets is the recognition that rural Alaskans depend on the fish and wildlife on these lands for sustenance, as well as for other important cultural, economic, and resource values. Therefore, subsistence hunting by qualified individuals is allowed in almost all ANILCA parks.

including the natural abundance, diversity, and genetic and ecological integrity of the plant and animal species native to those ecosystems."[4]

In accordance with the NPS mission, parks function as wildlife sanctuaries, aiding in the conservation and restoration of wildlife populations, including brown bears. In the Lower 48 states, Yellowstone, Grand Teton, and Glacier national parks, along with Waterton National Park in Canada, are core conservation areas for brown bear recovery. Brown bears also occur in North Cascades National Park, but the range of this small and isolated population is not contiguous with the other recovery areas; thus, efforts there have focused on habitat protection. Because of the wide-ranging nature of brown bears, recovery is contingent upon the ability of bear populations to expand to adjacent areas. This has resulted in conflicts with other landowners as development has increased and parks have become more isolated.

MANAGING FOR BEARS IN ALASKA

The current status of brown bears in Alaska differs from that of the Lower 48 states.

According to the Alaska Department of Fish and Game, Alaska is home to an estimated 30,000 brown bears.[5] Because brown bears are found widely across Alaska, protection under the ESA has been unnecessary here. As a large and relatively undeveloped state, Alaska has vast expanses of largely intact ecosystems. Parks and other undeveloped lands provide what brown bears need to thrive while also providing for human uses, including hunting and wildlife viewing.

Brown bear encounters are relatively common when people explore Alaska. In fact, viewing bears in their natural habitat is one of the primary reasons people from across the world travel to Alaska. And Alaska's national parks provide outstanding viewing opportunities. Bear viewing has continued to increase along the salmon streams of Katmai, in coastal sedge meadows of Lake Clark, and on the tundra of Denali.[6] For many out-of-state visitors, the opportunity to harvest or photograph a brown bear in Alaska is a once-in-a-lifetime experience. For Alaska Natives, bears have an immense cultural value that is part of their history, everyday life, and community. The connections of visitors and residents to bears highlight the social, cultural, traditional, nutritional, and economic value of bears and

Wildlife biologist Grant Hilderbrand and pilot Troy Cambier weigh a tranquilized bear in Lake Clark National Park and Preserve. (NPS)

reflect some of the experiences the NPS endeavors to provide in perpetuity.

The NPS views brown bears and humans, and their interactions, as part of the natural ecosystem; we, as NPS stewards, are charged with preserving both the ecosystem and the natural processes therein. Because each park is different, we rely on understanding the specific needs of local bear populations and the potential effects of a variety of human activities and developments.

ASKING THE RIGHT QUESTIONS

Almost no North American bear population has been studied in detail over a long period of time. One notable exception is the Yellowstone grizzly bear population. Due to the 1975 listing of this population in the Lower 48 states under the 1973 Endangered Species Act,[7] this ongoing research has required a collaborative and collective pool of expertise and resources from scientists and managers from numerous agencies, including the United States Fish and Wildlife Service, United States Geological Survey, NPS, United States Forest Service, and the states of Wyoming, Montana, and Idaho. Even with decades of effort, more is being learned, the environment continues to change, and technologies for studying bears advance. In contrast to Yellowstone bears, most bear populations in Alaska have either never been studied or have been studied for a limited time with a focus on specific management questions.

Based on what we know about bear ecology and bears' relationship to their environment, we can ask myriad questions to better understand the factors that benefit or harm local bear populations. Despite all these important unanswered questions, time, logistics, and financial resources are limited. How do we, as biologists, determine which questions to pursue when designing scientific studies of brown bears on national parklands in Alaska? Most studies originate as a question from park managers, such as, *how will this* [activity, change, or decision] *affect bears or the natural function of the ecosystem*?

The types of decisions we, as the NPS, may face range widely, from direct human impacts to indirect effects of development to understanding ongoing ramifications of climate change. Approximately 70% of all NPS lands in Alaska are open to some form of hunting (subsistence or sport hunting or both).[8] In addition, brown bears are not restricted by jurisdictional boundaries and commonly use both parklands and adjacent lands. Thus, they may be subject to either federal or state harvest regulations or both. In general, when managing harvest (or any kind of human-caused mortality), we focus efforts on trying to understand the demographics of the population—the number of bears in a particular place at a particular time, birth rate, and survival.

An additional area of inquiry often relates to human development, both in and adjacent to parks. Specifically, developments of interest in Alaska include park infrastructure, roads, and future potential mineral exploration and extraction. The fundamental questions we as researchers try to address related to development include potential effects of increased access to remote areas (most of Alaska is roadless), how animals move about and use the landscape, and which critical resources that bears rely on (e.g., denning habitat, food resources) will be impacted. Studies of bear habitat use and diet most often require us to collar a subset of the population to track them. Modern GPS (Global Positioning System) collars can collect bear locations every few minutes and allow us to analyze the habitats bears select (or avoid), important routes of travel, concentrated locations of important food resources, and den locations—denning is a particularly vulnerable time in their life. Detailed dietary and physiological studies similarly require that we handle animals to collect information on body weight, body fat, and seasonal diet by collecting and analyzing hair and blood samples. This approach gives us insights into the health of individual animals and helps refine our understanding of what resources bears are using on the landscape. Collectively, we can understand where bears are, why they are there, and what foods or habitat types are critical for the ongoing health of the population.

Often, wildlife management is, in fact, the management of people. Parklands in Alaska are used by local rural residents for subsistence hunting and gathering and for their cultural values. In addition, visitors from all over the world are drawn to parks to hike, camp, take photos, climb, view wildlife and vistas, hunt, fish, and simply enjoy nature. Our primary goal as park managers, relative to human-

bear interactions, is to keep both bears and people safe and to minimize stress and disturbance to bears and all wildlife. Our research into human-bear interactions often focuses on areas where bears and people commonly encounter one another, specifically bear-viewing sites (e.g., Brooks Camp in Katmai National Park and Preserve), the Denali Park Road, and campgrounds and trails. Study designs are often very specific to the location and information needs but may include collaring (e.g., the Denali Park Road study; see chapter 8), observational work (e.g., Brooks Camp; see chapter 10), or social-science surveys. The goals of these types of studies often include improving and refining our education and outreach messages to visitors to minimize adverse interactions between bears and people and promote the safety of both.

Another major topic of interest for park managers, and for our global society, is the impact of climate change. Like us, bears live in dynamic environments. Parklands in Alaska go through seasonal extremes; and bears are highly adapted to changes in weather, precipitation, and temperature. Their reproductive cycle and period of fat accumulation (hyperphagia) in the fall are evolved to match periods of food availability—abundance as well as scarcity—to support both themselves and their offspring through a period of winter dormancy. Because bears are so well adapted to their environment, they may be vulnerable to environmental change. Prior and current work provides baselines that we can use to assess the effects of climate change in future studies. By understanding how adaptive individuals can be in behavior, diet, and habitat selection (a concept known as *plasticity*), we can infer how adaptable bear populations are and how resilient they are to change. While some climate changes may be subtle (e.g., shifts in the distribution and timing of berry ripening), others may be extremely obvious and hugely impactful (e.g., heat stress on Pacific salmon that decreases salmon abundance or glacial retreat that creates new habitat).

While studies are often designed to inform a single or specific question or action, we also collect information opportunistically, as a cost-effective way to learn more about basic brown bear ecology and develop baselines for future studies. As an example, if bear capture and handling is part of a habitat use study, additional samples related to genetics, contaminants, stress, and diet can be collected at no additional cost. For non-invasive studies, such as hair-snare–based genetic population estimates, more detailed assessments of genetic relatedness and diet can also be conducted using the same samples. As these techniques are employed across studies, it allows us to develop baselines for a particular park and comparison across parks. Efforts such as these help us interpret park-specific findings and contribute to our broader understanding of bears as a species and of the environments in which they live.

THE PATH AHEAD

This book describes how we, as the NPS, conduct natural resource research and management more broadly, especially for brown bears. In the following chapters, we describe the relationships between bears and people, bear ecology, what we're learning about how bears adapt to climate change, bear research in parks, and how our knowledge is applied to park management. This book is not meant to be the last word on brown bears in Alaska. Rather, it captures a moment in time in a millennia-old relationship that, we hope, will continue for millennia to come.

NOTES

1. United States Congress, National Park Service Organic Act, 1–4.
2. United States Department of the Interior, National Park Service, *Management Policies 2006*, chapter 4, "Natural Resource Management."
3. United States Congress, Alaska National Interest Lands Conservation Act.
4. United States Department of the Interior, National Park Service, *Management Policies 2006*, chapter 4.
5. Alaska Department of Fish and Game, "Species Profile: Brown Bear."
6. National Park Service, "Visitor Use, Katmai National Park and Preserve."
7. White, Gunther, and van Manen, *Yellowstone Grizzly Bears*.
8. United States Congress, Alaska National Interest Lands Conservation Act.

Bears in Alaska live in dynamic environments. Our research is intended to expand our knowledge so we can conserve bears and their habitats within the context of human activity and climate change. (NPS/Tania Lewis)

Posing with their family bear spear, last used by their grandfather Roosevelt Jon, are (*left to right*), Stanley Starr, Al Starr Jr., and Randy Starr. The spear was used traditionally in winter hunting. Hunters would wake the bear in its den and brace the spear at the den opening so the charging bear would fall onto the point. Denali National Park and Preserve, June 2023. (NPS)

2 BEARS AND HUMANS

AN ANCIENT RELATIONSHIP THAT PERSISTS TODAY

Rachel Mason, Amy Craver, Dael Devenport, Karen Evanoff, Victoria Florey,
Mary Beth Moss, Marcy Okada, Jack Omelak, and Dillon Patterson

Bears and humans have coexisted in Alaska for thousands of years, and Indigenous people and settlers have used bear for meat, fur, and tools. People have accorded cultural and spiritual significance to bears as well and have developed rules for respectful handling of them. Brown, black, and polar bears all live in Alaska; this chapter focuses on brown bears.

Here we provide an overview of how different cultures use bears for subsistence, elaborate on ritual treatment of bears, discuss archaeological evidence of brown bear use in Alaska, and offer several contemporary case studies from different parts of Alaska—providing a closer look at the relationship between people and bears. The distribution of brown bears and Alaska Native groups overlaps extensively. Obviously, interaction between people and bears was unavoidable historically, and it stands to reason that bears would be well integrated into Alaska Native culture.

Several themes within our cultural understandings of bears exist across Alaska. Bears are large, powerful animals that are less important as a food source but more respected for the role they play in the

https://doi.org/10.5876/9781646427116.c002

spiritual world. Bears are often perceived as closer to humans than other animals, and brown bears are usually seen as more physically and spiritually powerful than the smaller black bears and closer to human beings in the spiritual world.

TRADITIONAL BEAR HUNTING PRACTICES AND RITUALS

Koyukon Athabascans' reverence for the bear's powerful spirit is expressed in the *hutlaanee* (taboo) system.[1] It is considered *hutlaanee* to say the word *bear* in the Koyukon language, and hunters show respect for the animal by not referring to it directly but instead using language that could be equated to a riddle. Instead of saying the word *bear*, Koyukon Athabascans use the phrase *big animal*. It is bad manners to look at or point to a bear.[2]

Many rituals and taboos are attached to bear hunting, both in preparation for the hunt and in the treatment of the bear during and after it. Bears are understood to have a special ability to detect the presence of hunters and assess their worthiness. Because bears are so dangerous and powerful, hunters need to take ritual precautions and treat the bear with respect. Several Alaska Native groups have proscriptions on talking about bears at any time but particularly before and during a hunt.

Tlingit bear hunters traditionally bathe, fast, and abstain from sex prior to a hunt. While in the field, hunters do not eat, drink, or build fires. Back at home, a hunter's family tries not to move around too much or become angry with others, as these behaviors may impact the success of the bear hunt. Some offenses, such as laughing at or ridiculing a bear, may provoke the bear to attack or act aggressively toward hunters.[3] Athabascan, Iñupiaq, and other Alaska Native hunters have traditionally adhered to similar proscriptions around bear hunting.

Until firearms became more widely used in the late nineteenth century, Alaska Natives' usual weapon for bear hunting was a spear. For Athabascans living in what is now Denali National Park and Preserve, the winter hunting method involved awakening the bear in its den and planting a spear in the ground in front of the den opening. Instead of throwing a bear spear, the hunter held it with the shaft end braced in the ground so the charging bear would "fall" onto the point. Hunters moved quietly toward the den where the bear could be seen sleeping. If the bear could not be seen through the main entrance, the hunter blocked the exit with brush and then cut into the side of the den and used an arrow or a spear to impale the bear.

In the spring, after bears emerged from their dens, Athabascan hunters used ground squirrels to attract bears. The hunter would release a squirrel near the bear. When the bear went after the squirrel, the hunter threw a spear at the bear. Dena'ina

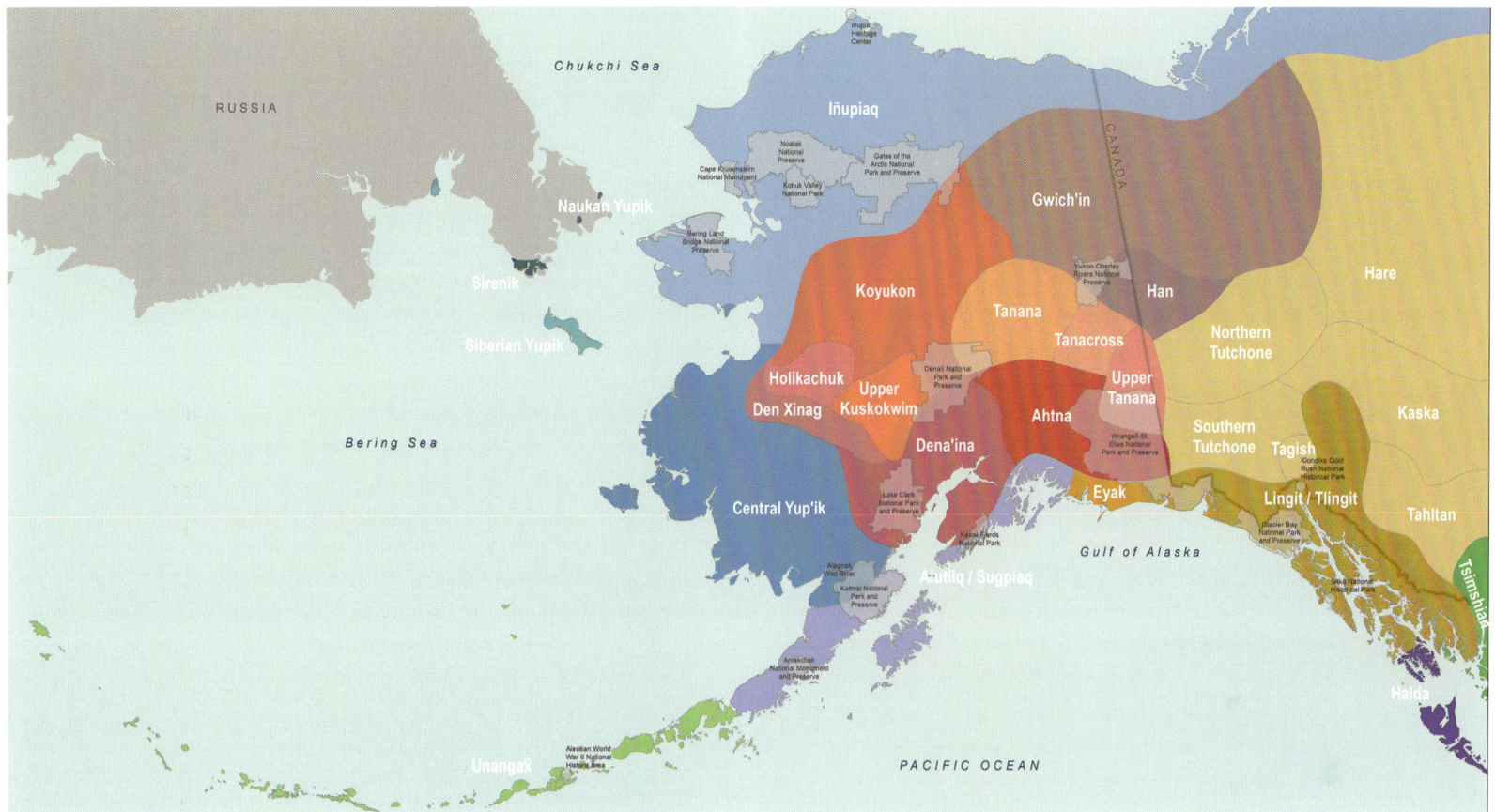

The geography of Indigenous peoples of Alaska overlaps with brown bear habitat. Throughout Alaska's history, the interaction between brown bears and people would have been unavoidable. Because of this, bears are well integrated into Alaska Native culture. (From Krauss et al., *Indigenous Peoples and Languages of Alaska*)

Athabascans used a spear tied to a long pole, bows and arrows, or a deadfall trap. They also set a snare for bears by digging a hole between 2 trees and putting bait above the hole. Another Dena'ina method was to hunt bears from a scaffold high in the trees along the river, when the big animals came out at night to fish.

For the Tlingit of southeast Alaska, the most common traditional hunting method was to locate a bear in its den, when it was most defenseless. Dogs were

Dena'ina Athabascan, Yup'ik, Iñupiaq, and Tlingit Use of Bears

—Karen Evanoff

Dena'ina Athabascans live in south-central Alaska and on the Alaska Peninsula. The Dena'ina communities of Nondalton, Pedro Bay, and Lime Village are associated with Lake Clark National Park and Preserve.

There was a time when the Dena'ina people spoke a language with the animals. Elders of the Lake Clark region have a close intrinsic relationship and connection with the natural world, offering far more than the intellectual brain can understand. There is no way anyone can live close to nature for a long period of time and not sense a deeper intelligence. In the Dena'ina language, there are words for this: *q'et ni' yi,* meaning "it is speaking to us," referring to the natural world, and *no-cultoni,* "the breath of life," leading to the relationship with big animals, such as bears.

What I share here is what I learned growing up and what I learned from Elders. This moral code of conduct guides every aspect of maintaining a good relationship of equality and respect. What I write here is so much bigger than me, bigger than all of us, and it is important to recognize the limitations of the written word that can, in fact, minimize a culture's thousands of years of existence.

Bears were regarded with such high respect that the animal's direct name was not used; rather, people referred to them as *big animal* to prevent any offense or disregard to the big animal. Good things will come to you in return if you show proper respect. When a bear was taken (killed), it was very important to cut out the eyes and bury them so the spirit could not see what is happening; again, this is a sign of respect. Big animals very closely resemble human beings when they stand, another reason for the kinship and high regard.

Long ago, people ate a lot of bears because moose and caribou were less plentiful. The meat was good eating, and the fat was rendered for oil because it does not freeze solid during the winter. The intestines were dried and used to make rain jackets. The stomach lining was dried and used as a sack. The animal's hip was used as a big spoon. Birch baskets were lined with dried bear stomach and the fat was stored in them. Bear fur was used was for blankets, parkas, and mittens.

trained to sniff out bears and flush them out of the den. When it emerged, the bear was dispatched with a spear. Alternatively, the bear was speared in its lair from a hole made in the side of the den. The Tlingit also caught bears in snares, deadfalls, and traps.[4]

The Tlingit consider bear hunting a very dangerous pursuit; successfully harvesting a bear affords the hunter high social prestige within the community. Other Alaska Native groups also consider bear hunters an elite group of skilled hunters. Traditionally, Yup'ik communities in southwest Alaska identify a few men as expert bear hunters who hold the knowledge required to prepare for and execute successful hunts according to customary rites.[5]

Many of the rituals connected with bear hunting involve respectful treatment of the bear after it is

In 18 Century men was killing bear by Spear it might be no easy but no way to kill the bear beside bear spear & Bow & arrows spearhead are made out of front leg bone of the Grizzly bear in north Slope. Continued about catching bear,

Simon Paneak
1968

killed. Athabascan hunters say the bear gives itself to the hunter; in return, they respectfully thank the bear for allowing itself to be taken. Before processing the bear, they cut out both its eyes so the bear does not see the hunters. The small tendon under the throat is also removed to prevent the bear from speaking to other bears about the hunters. Elders from Nikolai say that in the past, after the bear was butchered, hunters put the bear's head on a short tree so the bear could rejoin its relatives. Sometimes a small tuft of fur from the bear was tied to the spear as a measure of respect for the animal.

Yup'ik hunters in southwest Alaska emphasize cleanliness in bear hunting. Bears have a keen sense of smell, making less-clean hunters easier to detect. Hunters must also avoid talking about bears they

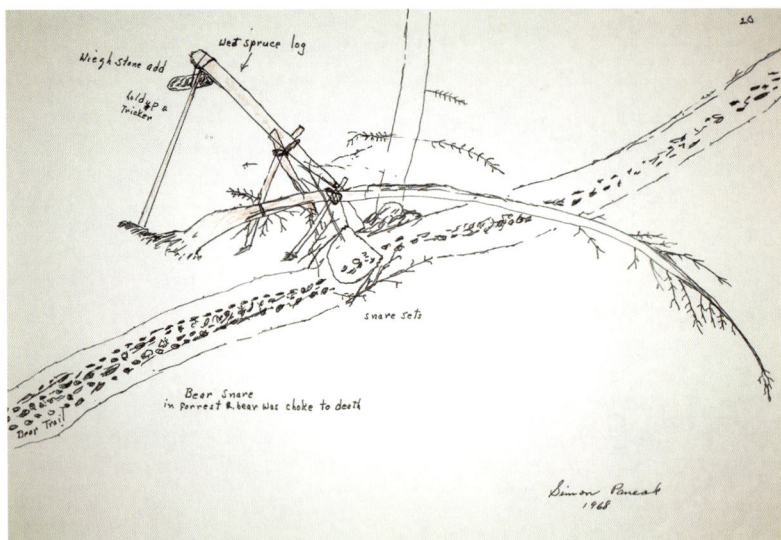

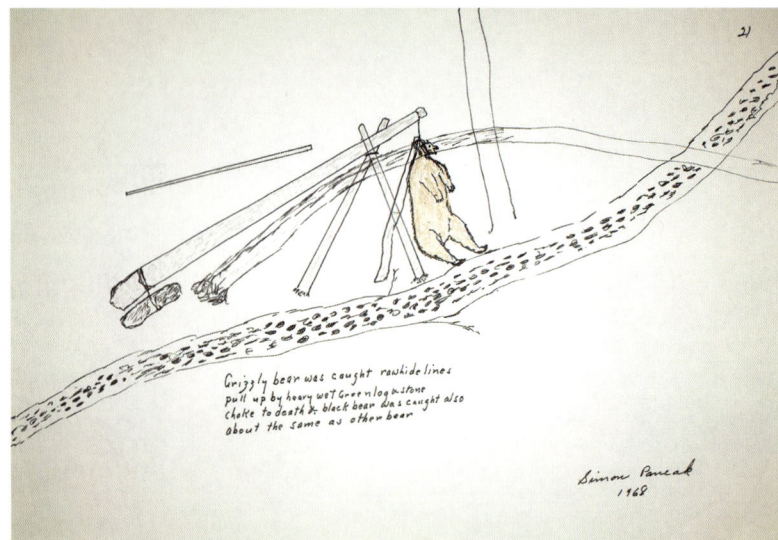

Diagram showing a snare made to capture bears. A bear could be caught around the neck by the rawhide lines of the snare, which were pulled up and tightened by the release of a green tree trunk weighted with a stone. This method was used for both brown bears and black bears. (Simon Paneak, John M. Campbell Papers, GAAR-00219, National Parks Service office. Simon Paneak, Nunamiut hunter from Anaktuvuk Pass, Fairbanks, Alaska)

intend to hunt, as bears might listen and hear where a hunter is going to be.

Bears will then avoid those locations, making hunter success far less likely. While butchering the bear, practices such as cutting out the eyeballs and stuffing them in the nostrils, driving sticks into the ears, and slicing the flesh along the jaw ritually prevent the bear from seeing, smelling, hearing, and biting.[6]

A common practice in many traditions involves proper treatment of the bear's skull. Yup'ik hunters lay the skull in the field facing east, either buried beneath boulders or sunk in a body of water.[7] Iñupiaq hunters remove the hyoid bone from beneath the tongue. The hyoid bone is sometimes placed in willow branches or tussocks, other times buried or discarded. This practice is done to ensure the bear's spirit has left and will not harm the hunter. Iñupiaq hunters also leave the skull in the field, either at the harvest location or at camp—sometimes hanging in a tree, sometimes on the tundra facing west.[8]

WOMEN AND BEARS

Often, women must observe special norms and taboos regarding bears. For example, women in many Athabascan groups are forbidden to eat bear meat or have contact with bear hunters or hunting weapons, especially while the women are pregnant

or menstruating. Upper Kuskokwim and Koyukon people believe women of childbearing age should not look at a bear or eat bear meat, lest harm come to their unborn children. Hides are sometimes disposed of in the woods to keep women from stepping over or touching bear hides.[9] However, young women are taught how to process and preserve bear meat. In some areas, only older women who have gone through menopause may scrape and tan a bear hide.[10]

The Tlingit, in contrast, hold that women have special power over bears and that this power can be used in a bear hunt. A Tlingit informant told Frederica de Laguna: "When a woman shoots bears,

LEFT: Chilkat tunic with sea otter fur and leather fringe trim, owned by the Bear House, Chookaneidí Clan Tlingit. In Tlingit culture, bears and humans have a close relationship. (Ron Klein, Chookaneidí Clan)

RIGHT: Tlingit dancing bib depicting the bear, a Chookaneidí Clan crest. (Ron Klein, Chookaneidí Clan)

Xóots, Friend and Foe

—*Mary Beth Moss*

Glacier Bay National Park and Preserve is the traditional Homeland of the Huna Tlingit.

On a blustery summer day almost 30 years ago, a group of Tlingit Elders accompanied by National Park Service employees climbed out of rubber dinghies on the shores of Dundas Bay in Glacier Bay National Park and Preserve. We were there to harvest berries—nagoon berries, to be exact. These small, flavorful raspberry-like gems are treasured by the Huna Tlingit. Park staff were there to assist, to learn, and to ensure the safety of Tribal members in a land that hosts numerous bears. The Huna women were not impressed with the bear spray canisters the park staff carried. They instead held out soda cans filled with stones from the beach, gently rattling them as they approached the beach-grass fringing the shoreline. "*Eesháan xat,*" they called. "*Pity me. We are only here to pick a few berries. We will leave many for you. Have pity on us, brothers.*" The Elders told us: "*You know, they can hear you. If they're near you, they'll just walk away.*"[11] Their entreaties were clearly received; we encountered no bears that day, and the Elders filled their buckets with sweet berries.

"*Clan crests,*" Chookaneidí Clan matriarch Lily White told me, "*come from moments of peril*"; for certain Tlingit clans, the brown bear (*Xóots* in Tlingit) is a primary clan crest.

Stories of *Xóots* are tightly woven throughout the history of the Chookaneidí Clan, whose sacred place of origin is Chookanhéeni (Grassy River) in what is now Glacier Bay National Park and Preserve. Bear stories link one phase of history to another.

Archaeological evidence and oral history suggest that the ancestors of the Chookaneidí Clan once lived at Xákwnoowú, a fort and village site in present-day Dundas Bay, a large fjord just west of Glacier Bay proper. Xákwnoowú appears to have been left unoccupied for periods of time, possibly as a result of an epidemic alluded to in various oral histories. Two such histories feature brown bears and likely memorialize both the tragic loss of an entire village and the resulting movement of a handful of survivors to a settlement site inside Glacier Bay proper.

Though the stories differ in detail, they both describe a man who, lonely and grieving the loss of his family, invites brown bears to the required memorial feast for the dead. After feasting on berries and other delicacies, the chief of the bears directs his brethren, "Do not leave this man friendless, but go to him every one of you and show your respect."[12] From that encounter on, "When they gave a feast, no matter if a person were their enemy, they would invite him and become friends just as this man did to the bears."[13]

Although the stories leave off there, we know that generations of Chookaneidí, as well as other clans, flourished in their new settlement in S'é Shuyee (area at the edge of the glacial silt, Glacier Bay), an area rich with salmon streams and berries. Numerous oral histories describe the area as a veritable Garden of Eden, with many clan houses lining the shores.

But the Chookaneidí would face many more hardships in years to come. Lily White's narration of her clan's history recounts an arduous journey from S'é Shuyee Homeland to the protected shores of Xóots Geiyí (Bear Bay, now Port Frederick) during the glacial advances of the Little Ice Age. An especially harsh period—Wooshu T'aakw (the Year of Two Winters)—worsened an already treacherous situation. When summer failed to bring its riches that year, a Chookaneidí shaman used his powers to feed the starving clan, but their kin, the brown bear, also aided the survival. From a small peephole in the clan's fortressed position in Xóots Geiyí, a man witnessed bears—hundreds of bears—along the nearby shoreline performing strange tasks:

All the way, they stand. They dug a big ditch. He said they were slamming something and they were eating it. He was watching them. When they finally left, he woke up everybody. He said "they were eating something. They dug up that whole beach." Big ditch. He told them to take their baskets, "Let's go look." When they went to look, they were real big cockles, clams, mixed. They were just picking them. "Wash it first," he told them. Full of sand. They washed it in salt water. They took them home and put them on short sticks like barbeque. They were eating those cockles. They said the bears taught them how to eat cockles.[21]

Despite the harsh climate, diseases, dense bear populations, and ensuing wars, the Chookaneidí Clan eventually prospered in Xuniyaa (Hoonah), building first Naa Naa Hít (Up River House), then Xáay Hít (Cedar House), then Xáatl Hít (Ice House), and then Xóots Jini Hít (Bear Paw House).[15] This last house drew its name from an epic tale of a marauding bear. Chookaneidí clansmen had made several failed attempts to kill a large bear ravaging the village, but the bear's strong rib cage ("*like brass*") prevented a knife or spear from entering it. Lily White tells how this particular "s'uk̲kasdúk̲," closed-rib bear, was finally overcome:

> Nothing but bear caves up here [cliffs above Hoonah]. There was a man eater among them. Just went after men and eats

men. So they got ready. Any one of them try to get in there, it just kills them. That one of them got ready. He had them tie fur to his body on the back. When the bear came at him, when he went into the bear cave. He went in backwards. When the bear got him it didn't touch him, he just turned around and got him right up here [on the head]. Real big bear. He killed it. It didn't go after animals or anything, it went after the men. That's the history, that's how come Chookaneidí have bear design.[16]

The hero in this story, Elder Shirley Kendall once told me, was her grandfather, James St. Clair, the Chookaneidí Clan leader at the time. Kendall noted that her family believes the incident actually occurred in Chookan-héeni, Glacier Bay:

> And he wasn't a very big man, he was built like my dad was, very small. Apparently, there was a brown bear that became rogue and began to kill people . . . And the women even became afraid because they would take their children to go berry picking and they couldn't go. So they, somebody, I'm not sure who, discovered the bear den . . . Then apparently my grandfather had a knife, a spear, he had a spear. He went into the cave, he ran in there. He went all the way to the back of the cave and there was a big rock close to the wall. He got behind it and when

his eyes got adjusted to the darkness, he could see the bear laying there, watching him . . . So, all he had was the spear . . . The bear stood up and walked over and looked him over. Then he attacked. My grandfather speared him into the chest and the bear pushed it so hard against the wall that it broke the spear. And so the big struggle was for him to pull the spear out again and to use it, but the bear was attacking him so he couldn't get a good grip on it. In the middle of all that big fight this lady was still out there antagonizing everybody and she said, "Nobody is going to go in there to help? He's the only one that's brave!" Her son stood up and he went in, and all he had was a knife. When he went in, he jumped on the bear's back, put his arm around that bear's neck. So, the bear tried to reach him, which gave my grandfather the opportunity to pull that spear out. So, he was using the spear from the front and the young man was pushing the knife into the neck of the bear. So, every time he attacked the bear, the bear tried to reach him [the boy] and my grandfather would use the spear. Together they killed him. But my dad said that his father went blind at a later age because the bear in its last stages was breathing really hard and his saliva was getting into his eyes. Once he got out of there, it was hard to clean him out. It eventually destroyed his eyesight.[17]

Xóots, Friend and Foe (continued)

St. Clair and his family had returned to their remembered but much changed ancestral homeland, now called Sít' Eeti Geiyí (Bay in Place of the Glacier, Glacier Bay) almost as soon as the ice retreated and salmon recolonized the streams. Perhaps as early as 1850, St. Clair established Xóots Hít, Bear House, in Chookanhéeni, commemorating his victory over the brown bear. *Xóots*, essentially, brought the Chookaneidí full circle, back to Glacier Bay.

Many years after the berry-gathering trip in Dundas Bay, the Chookaneidí Clan came together on the back deck of a tour boat with the Kaagwaantaan, T'akdeintaan, and Wooshkeetan Clans who also call Glacier Bay their homeland. Floating in front of S'it Tlein (Margerie Glacier in Glacier Bay), they sang mourning songs and ritually fed the ancestor left behind in the ice, as they did most years. Afterward, the boat was still as everyone listened and watched for the anticipated response; sometimes a glacier groaned or creaked, occasionally an iceberg calved noisily into the sea. On this day, Tlingit Elders and park employees alike heard distant drumming and singing floating across the bay. And in the center of S'it Tlein, a darkened portion of the glacier shaped exactly like a huge brown bear appeared. For the Chookaneidí, *Xóots* is both friend and foe.

---

the bears can't do no harm. If a woman cleans your gun, then the bear knows and he just drops . . . It's just like they [the women] make a wish, I think. One shot and they [the bears] just drop. When it comes like that, the bear just got no power."[18]

The Tlingit believe that if a bear observes an unclothed woman, it will become embarrassed and retreat.[19] In the days when women wore large labrets (lip plugs), if a woman met a grizzly bear, she could take out her labret and blow toward the bear through the hole of her lip, and the bear would not touch her.[20] Another way to drive away a meddlesome bear was for a human widow or pubescent girl to grab mud from bear tracks and throw it into boiling water.[21] An adolescent girl could turn a bear to stone by looking at it.[22] In the early 1990s, Thomas Thornton noted that Tlingit women were still able to influence bears by speaking to them in Tlingit or undertaking certain ritual actions.[23]

In Tlingit culture, humans and bears share a close relationship, founded, in part, on their similar skeletal structure. This relationship is manifested in various stories in which humans and bears transform, marry, or both. In the Tlingit story "The Woman Who Married the Bear," a young woman picking berries steps in bear excrement and curses the bear. A bear appears to her in the form of a young man, and the pair fall in love and marry. The woman's brothers eventually find her in the bear settlement and kill the bear. The woman briefly returns to the human world but cannot adjust, preferring to live with the bear society in the mountains.[24] Another

Tlingit story tells of a man who marries a bear. In these stories, humans and bears live in parallel worlds, but some individuals can move from one to the other. Women are particularly able to make this connection.

Bears are often associated with human kinship groups. A number of Tlingit clans and houses claim the bear as a crest, a symbol of their identity. Members of these clans have a special bond with bears. The Teiḵweidí, Naanya.ayí, Kaagwaantaan, Sik'naẋ.ádi, and Chookaneidí are among those clans that claim the brown bear as a crest.

BROWN BEARS AND ARCHAEOLOGY IN ALASKA

Archaeologists have found evidence of humans using bears as a subsistence resource as well as evidence showing great reverence for them in sites around Alaska. The evidence includes animal remains demonstrating use of bears for food, fur, and tools and representation in artwork such as rock art or carvings of bears. Differential treatment of bear remains, and modified canine teeth have been documented in archaeological sites around the state, from the northwest coast to the southern border of southeast Alaska.

Hunting bears in Alaska has been documented through bones with cut marks or bones that had been modified for use as tools.[25] Bear blood residue on stone tools from a site in the De Long Mountains in northwest Alaska confirms that bears were used as a subsistence resource.[26] Sites dating to more than 10,000 years old in caves on the northwest coast of North America, including 1 in Alaska, may represent the traditional technique of hunting bears in their dens. Broken projectile points were found with bear remains in these caves.[27]

The differential treatment of bear bones that has been documented at some sites is evidence that bears were highly respected in several Alaska Native cultures. At a site on Haida Gwaii, an island 30 miles off the southern border of Alaska dating to 9,400 years ago, cut marks and a greater number of certain skeletal elements suggest ritual practices. On Dundas Island, 5 miles south of the Alaskan border, an ochre-marked bear skull was found on a midden

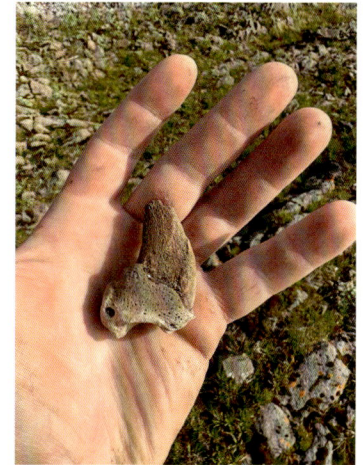

LEFT: A grizzly bear metatarsal/metacarpal bone from a cave in Yukon-Charley Rivers National Preserve dating to more than 42,000 years BP. (NPS/Jeff Rasic)

RIGHT: This bear tooth was found in a midden—an ancient trash heap filled with the remains of hunted and cooked animals—from a coastal Alutiiq village occupied between about 1,100–5,500 years BP. The village was located on an island with access to the Shelikof Strait, between the Alaska Peninsula and Kodiak Island. (NPS/Sebastian Wetherbee)

A chert projectile point that reportedly has microscopic traces of bear residue. Collected from the North Slope of Alaska near Gates of the Arctic National Park and Preserve. Similar artifacts occur in multiple parks. (University of Alaska Museum collection, UA78-80-633, Jeff Rasic photo)

with a wooden bear carving placed in a nearby rock crevice.[28] A charred brown bear scapula found on Kodiak Island aligns with ethnographic accounts of burning bear bones as a form of respect.[29]

Brown bear remains appear in archaeological sites all over Alaska associated with different cultures; however, faunal analyses show that bear remains usually make up a small percentage of the total, usually less than 4%.[30] One reason for the relative rarity of bear remains could be that brown bears were not a primary subsistence resource like salmon or seal. The paucity of remains may also be a result of cultural practices regarding the respectful treatment of bear remains, such as burning or burying bones.[31]

Many Alaska Native peoples rely heavily on salmon as a primary subsistence resource, and settlements are frequently located along salmon-rich waterways. Bears are also attracted to salmon, yet they try to avoid humans whenever possible. The deterrence of human presence and their hunting practices may have kept bear populations lower than they are today in areas where humans do not currently live and where hunting is prohibited or seasonally restricted. One illustration of this is Brooks Falls in Katmai National Park and Preserve, which currently attracts many bears during the salmon run and bear-viewing visitors from all over the world. Yet, archaeological evidence shows that humans lived along the Brooks River for the past 5,000 years and may have prevented bears from accessing this resource as easily as they do today.[32]

Archaeologists on Kodiak Island have used other lines of evidence to provide information about pre-contact bear-human interactions. Surveys found that in salmon-rich areas, over half of archaeological sites had been damaged by bears. The researchers observed that younger sites are more frequently damaged by bears than older sites. Bears are attracted to the fresh vegetation that appears on archaeological sites early in the spring. The more recently abandoned sites are greener than the older ones due to the organic material still present, which supports vegetation growth.

People living nearby probably discouraged bears from accessing the older sites while they supported vegetation growth prior to organic decay. Now, sites are protected in parks and preserves, and bears have unrestricted access to the vegetation growing on the younger sites.[33]

Understanding how the original inhabitants of Alaska used bears as a subsistence resource and treated them with great reverence benefits archaeological research and helps us understand how historical and contemporary peoples have viewed and treated bears.

Brown bears continue to have immense cultural significance and to represent an integral part of Alaska Native subsistence lifestyles across the state. Many of the cultural beliefs and practices described above are still followed today. However, the extent to which brown bears comprise part of local subsistence economies varies tremendously from region to region.

USE OF BEARS IN THE PRESENT DAY

In Alaska, the word *subsistence* carries multiple meanings. Subsistence is, in the most literal sense, how people obtain food. Subsistence in Alaska is primarily characterized by hunting and fishing for wild resources, although gathering wild plants, berries, and mushrooms makes up an important part of subsistence diets as well. To that end, the term *subsistence* also has legal and regulatory meaning. Both the Alaska National Interest Lands Conservation Act, signed into law in 1980, and Alaska state law prioritize subsistence over all other consumptive uses, including sport and commercial harvest. When resources are scarce, resource managers are legally required to prioritize subsistence. Most of Alaska's national parks, preserves, and monuments are open to subsistence hunting, fishing, and gathering for people who permanently reside in rural communities where residents have customarily and traditionally engaged in subsistence uses.

While both Alaska Natives and non-Native rural residents are eligible to harvest wild foods for subsistence, the term *subsistence* has also been used by some Alaska Native communities to refer to traditional lifeways based on seasonal harvest of wild resources, steeped in complex cultural beliefs and practices. Other Alaska Native communities find subsistence an inadequate word to describe their way of life.

For Upper Kuskokwim Athabascans living near Denali National Park and Preserve, cultural taboos restrict consumption of bears. They are mainly eaten during potlatches and other ceremonial occasions and are used to feed dogs. Older residents

Norms around Bear Hunting

—Marcy Okada

Koyukon Athabascans are a subgroup of Athabascan Indians whose name comes from the Yukon and Koyukuk Rivers where their villages are found. There are eleven Koyukon Athabascan villages in Alaska. This case study focuses on the communities of Huslia, Hughes, and Allakaket, which are affiliated with Gates of the Arctic National Park and Preserve. The southern Brooks Range Mountains are considered the traditional homelands of the Koyukon Athabascan people.

I had the opportunity to interview Orville Huntington, originally from the Koyukon Athabascan community of Huslia, a place where cultural traditions hold strong and spiritual beliefs about bears continue. Orville said there is an unspoken component of learning and that spiritual beliefs are interwoven into experiential lifetime learning.

Knowing that it is considered *hutlaanee* (taboo) to talk about the big animal with women, before we started our conversation I asked Orville whether he felt comfortable having this discussion. Orville thought it would be all right since I don't come from the same cultural background. Orville shared the following information:

In regard to controlling brown bear populations, we don't like to kill animals for no reason, unless they are threatening our safety. There are what we call crazy bears or "red-eye" bears. We don't mess with these bears because they don't have a natural thinking order and they are not good to eat. It is Koyukon Athabascan belief that most animals think about what they're doing. It seems that animals are losing respect for humans these days.

There are other signs that climate change is affecting the bears. They sleep more lightly in the fall time now, and they are coming out earlier in the springtime in areas that are warm. Bears will always continue to adapt to new foods, and bears that eat certain foods at a certain time of year taste a certain way. Mountain bears that eat ground squirrels taste better, as do black bears that eat grass roots. Fall brown bears that eat rotten fish don't taste good.

Between the communities of Huslia and Koyukuk, only certain people hunted brown bears. We are careful on who we let hunt, and it's something not many people wanted to do. Koyukon Athabascan knowledge in Huslia is [that] it's important to keep bears in check because they'll take over and then there will be too many bears. Dogs were never used to hunt bears—they were used as pack animals and to pull sleds. Dogs protected their owners and were taken to camp for protection.

Orville said he's been hunted by brown bears while out on the land. One instance was while he was walking through the woods when there was no wind. Suddenly, he heard wind and immediately knew a bear was following him. The bear made a loud noise, and Orville ran into an open meadow and waited it out. There is a belief that you can talk to bears in the Koyukon Athabascan language and tell them to leave, calm down, or go eat something else. By speaking in Koyukon in a certain tone of voice, you can make the bear understand because their animal language is close to the Koyukon language.

of Nikolai and Telida say that in the past, they rendered the bear fat into grease that was eaten with dried fish or meat during hunting trips. Bear grease was believed to have medicinal qualities and was sometimes applied to burns, sores, and rashes.[34]

Brown bear concentrations are very high along the Alaska Peninsula and in the broader southwest region of the state. In Yup'ik, -/Alutiiq, and Unangax̂ communities in this region, brown bears are significant components of subsistence economies, particularly in times when other species that are dietary staples (e.g., moose, caribou, and seals) are scarce.[35]

Hannah Loon and Susan Georgette surveyed Iñupiat communities of northwest Alaska's Kotzebue Sound region in 1989 for the Alaska Department of Fish and Game to understand the contemporary subsistence use of brown bears. Georgette later conducted a survey of residents of Deering and Shishmaref to learn about subsistence uses of brown bears in the Bering Land Bridge National Preserve. A common theme from both studies is that women are very much a part of bear-processing activities, if not in charge of them. Women cut, process, and store the meat, skin, and other body parts. Several women interviewed said they used bear fat either medicinally or in place of oil or butter. Examples of interactions between women and brown bears are used as cautionary tales or models of how to interact with living bears. Elders advised women to bare their breasts to make the bear run away if an encounter occurred. One local informant said bears like to steal children, so humans need to always be careful not to let children out of their sight when they are outside.[36]

Gordon Newlin from Noorvik said that in the past, oil made from bear fat was used for both cooking and medicine. He said, "We don't use it that much anymore since we have clinics in every village. That cuts down how much people are harvesting bears." Hannah Loon from Selawik said bears had become more of a nuisance in the Lower Kobuk coastal area than in the Upper Kobuk area. She added, "Bear fat is very important to people. Like bacon or seal oil."

Though the specific rituals and rules may differ, the theme of respect for bears is consistent throughout Alaska Native cultures that have bear-hunting traditions. A primary purpose of these practices is to ensure hunter safety and success. Conflicts between these practices and hunting regulations highlight challenges inherent in contemporary subsistence management. State and federal agencies have historically based regulations on sport hunting practices, centered on trophies. For example, current regulations for the general hunt mandate the removal of brown bear skulls from the field. This requirement renders illegal the practice of Alaska

Northwest Alaska Case Study

—Jack Omelak

Nome, where Jack Omelak grew up, is a hub community for Iñupiat groups on the Seward Peninsula as well as for Yup'ik villagers from St. Lawrence Island. A gold rush around 1900 brought a huge influx of miners and others to Nome from outside Alaska. Its population in 2020 was about 3,700 and was 60 percent Alaska Native.

When tasked with the opportunity to contribute to this book, which focuses on grizzly or brown bears, I immediately faced a dilemma. Why have I not been indoctrinated through oral histories to the significance of brown bears to us as Alaska Natives? Why is there very little mention of brown bears and their relationships to the people in our region's principal literature of anthropology? Perhaps such information had been lost, and I am sadly again the victim of modernization, colonialism, and culture loss.

Yes, I had grown up in the mountains of the Seward Peninsula trapping squirrels and gathering fish and berries during the summertime. I, along with my mother and sister, walked to our campsites, which were adjacent to camps of older and widowed Eskimo ladies who were also camping "way out in the country" to trap squirrels in the spring, catch fish in the summer, and pick berries in the fall. We trapped the mountainsides for most of each day. At night my sister, mother, and I would walk the few miles to have dinner at one of the other ladies' camps. Although kids were rarely allowed to be seen or heard inside, we would listen to stories while the adults drank tea and ate crackers until late. Rarely did we hear about grizzly bears, other than the occasional story that one got into so and so's camp and destroyed it.

Or we would hear that if a woman was ever in a bear encounter, she should take off her shirt and expose her breasts. According to the story, the bear will become too ashamed to attack a female human and will leave. Stories of brown bears were rare, let alone those about cultural taboos or processes for veneration and adoration of them. We would then walk back to camp in the dark, unarmed.

At one time, perhaps, there were deeper and more meaningful relationships between brown bears and Iñupiaq dwellers of the coast, but as our dependence on natural resources has decreased as a result of a global economy, perhaps the relationships between the 2 species have also changed over time. Perhaps the simplest answer here is that our coastal groups had, and continue to have, a rich cultural heritage surrounding polar bears as we shared environments with them and competed against them for our primary source of food (seals). Brown bears, an inland species, are less important to human subsistence in everyday life and in the spiritual world.

Native hunting traditions involving the treatment of bear skulls and places a burden on subsistence users who would not otherwise pack hides and skulls out of the field. Under a state subsistence permit for brown bear, however, salvage of the hide and skull is optional. Contrarily, while federal subsistence regulations require hunters to salvage the edible meat of brown bears, most state hunting regulations do not. The concept of salvaging the skull but not the meat is offensive to some subsistence hunters.[37]

Brown bears continue to hold cultural significance today. Although the extent varies greatly by culture and region, bear hunting remains an important subsistence activity among Alaska Natives. Brown bear conservation is an important part of the NPS mission to protect bears for generations to come. (NPS/Jim Pfeiffenberger)

Brown bear meat and hides have traditionally played an integral role in subsistence economies in many parts of the state. However, in some regions, brown bears are scarcely used, if at all. Today, brown bears are primarily harvested at the request of Elders. Taste for brown bear is also a matter of individual preference and is subject to seasonal variation in meat quality.[38]

Increasingly, bears are thought of less as a food source and more as competitors for food (e.g., fish, moose, and caribou). Traditional reverence and rites are giving way to notions of competition and predator control. Nonetheless, some hunters continue to pursue brown bears, observing traditions that have been passed down through the generations.[39] Brown bear conservation is paramount not only to the National Park Service mission of preserving natural and healthy populations for generations to come but also for the continuation of subsistence in Alaska.

NOTES

1. Nelson, *Make Prayers to the Raven.*
2. White, "Way of The Hunter."
3. Thornton, *Subsistence Use of Brown Bear in Southeast Alaska,* 63–64; de Laguna, *Under Mount Saint Elias,* 365, 364, 64.
4. Thornton, *Subsistence Use of Brown Bear in Southeast Alaska,* 63–64.
5. Van Daele et al., "Grizzlies, Eskimos, and Biologists," 141–152.
6. Fall and Hutchinson-Scarbrough, *Subsistence Uses of Brown Bears in Communities of Game Management Unit 9E.*
7. Coffing, *Subsistence Use of Brown Bear in Western Alaska.*
8. Loon and Georgette, *Contemporary Brown Bear Use in Northwest Alaska.*
9. Nelson, *Make Prayers to the Raven.*
10. Simon, "Customary and Traditional Use Worksheet."
11. White and Marvin, "Huna Clans and Marriage."
12. Swanton, *Tlingit Myths and Texts*; Dekinaak, *Origin of Iceberg House,* 52–53.
13. Kasank, *The Man Who Entertained the Bears,* 220–222.
14. White, "Huna Tlingit Place Names Recordings."
15. White and Marvin, "Huna Clans and Marriage."
16. White, "Huna Tlingit Place Names Recordings."
17. Kendall, "Interview."
18. de Laguna, *Under Mount Saint Elias,* 364.
19. Kamenskii, *Tlingit Indians of Alaska,* 75.
20. Swanton, "Social Condition, Beliefs, and Linguistic Relationships of Tlingit Indians," 455.
21. Olson, *Social Structure and Social Life of the Tlingit in Alaska,* 122.
22. de Laguna, *Under Mount Saint Elias.*
23. Thornton, *Subsistence Use of Brown Bear in Southeast Alaska,* 63–64.
24. Dauenhauer and Dauenhauer, *Haa Shuká, Our Ancestors.*
25. Ackerman, "Bluefish Cave," 511–513.
26. Gerlach et al., "Blood Protein Residues on Lithic Artifacts from Two Archaeological Sites."
27. McLaren et al., "Bear Hunting at the Pleistocene/Holocene Transition."
28. Germonpre and Hamalainen, "Fossil Bear Bones in the Belgian Upper Paleolithic."
29. Kopperl, "Cultural Complexity and Resource Intensification on Kodiak Island, Alaska."
30. Partlow, "Salmon Intensification and Changing Household Organization in the Kodiak Archipelago."
31. de Laguna, "The Atna of the Copper River, Alaska," 11–12.
32. Birkedal, "Ancient Hunters in the Alaskan Wilderness."
33. Saltonstall and Steffian, "Archaeological Survey in Eastern Alitak Bay."
34. Stokes, *Natural Resource Utilization of Four Upper Kuskokwim Communities.*
35. Behnke, *Subsistence Use of Brown Bear in the Bristol Bay Area.*
36. Loon and Georgette, *Contemporary Brown Bear Use in Northwest Alaska*; Georgette, "Brown Bears on the Northern Seward Peninsula, Alaska."
37. Van Daele et al., "Grizzlies, Eskimos, and Biologists"; Loon and Georgette, *Contemporary Brown Bear Use in Northwest Alaska.*
38. Jacobs and Jacobs, "Southeast Alaska Native Foods; Thornton," *Subsistence Use of Brown Bear in Southeast Alaska.*
39. Van Daele et al., "Grizzlies, Eskimos, and Biologists."

▸ Increasingly, bears are less often thought of as food than as competitors for food. While some hunters continue traditional hunting and use of bears, others see it as predator control. (Mark Johnson)

Bear behavior is largely driven by food. Their movements track the availability of highly nutritious food sources as they become available. (NPS/Jim Pfeiffenberger)

3 ECOLOGY

HOW BEARS LIVE IN THEIR WORLD

Grant V. Hilderbrand, Kyle Joly, William Deacy, and David D. Gustine

Alaska is one of the few places in North America with extensive brown bear habitat and healthy populations. Brown bears are 1 of 8 species of bears that exist in the world today; the others are Asiatic black bear, sun bear, American black bear, sloth bear, polar bear, Andean bear, and giant panda. Common names for brown bears in North America include grizzly bears (usually interior bears living away from the coast), brown bears (often referring to coastal brown bears), and Kodiak bears (those bears living on Kodiak Island). All of these brown bears are the same species (*Ursus arctos*), and we will refer to them collectively as brown bears. Brown bears presently occur on 3 continents: Europe, Asia, and North America. Historically, brown bears spanned much of the North American continent, ranging from central Mexico in the south to north of the Arctic Circle and eastward throughout the mountains and plains to the Mississippi River. Due to habitat loss, development, eradication efforts, and sport and commercial hunting, their reduced range is now limited to Alaska, western Canada, and relatively small, somewhat isolated populations in the Lower 48 states.[1]

https://doi.org/10.5876/9781646427116.c003

Current range of brown bears in North America (shaded areas). Historically, brown bears ranged most of North America, from central Mexico to north of the Arctic Circle. Now their range is reduced to Alaska, northern and western Canada, and small populations in the Lower 48 states due to purposeful overharvest, habitat loss, and development.

LIFE HISTORY

Life history is a description of the strategies an animal uses to grow, survive, reproduce, and generally optimize benefits and minimize risk from its environment. Brown bears have a complex and highly adapted life history. As biologists, we study life history to better understand brown bear population health and dynamics.

One fundamental way we assess the health of a population of animals is through their demographic statistics. The most common demographic statistic is the size of a population or number of animals. Since most bears live in expansive landscapes where they can be hard to see, almost all population numbers are estimates. Generally, estimating populations is expensive and labor-intensive, and it results in numbers with a large range in uncertainty. Consequently, populations are not estimated often, and few bear populations throughout Alaska have current estimates of population size—and never have. That said, techniques to estimate the number of bears in an area are evolving and improving (see chapter 6).

Without a specific population number, we use other important statistics to both understand the current status of a population and predict whether it will rise or fall in the future. These include annual survival rate (how many individuals survive the entire year), reproductive rate (how many cubs are born and survive their first year), average number of years between successive litters, age of first reproduction, and average age at death. In essence, if the number of cubs that survive to join the population as adults exceeds the number of deaths, the population will increase. The opposite results in a declining population. Evaluated together, these statistics help us understand the status and trend of bear populations and are needed to guide management decisions, such as setting an acceptable harvest rate.[2]

HIBERNATION AND REPRODUCTION

Hibernation (when bears go into their dens) is a life history strategy brown bears use to reduce energetic needs at a time of year when food is scarce. Although bears can be easily disturbed or roused during the denning period and can leave the den periodically during winter, this is generally a time of hibernation. To minimize the energy needed to survive this period of winter inactivity, both heart rate and respiration rate are greatly reduced. When active, bears maintain a body temperature similar to that of humans; however, this drops to around 90°F (32°C) during winter torpor. During their winter torpor, both heart rate and respiration rate drop substantially; however, because their body temperature remains relatively high, they are not true hibernators like Arctic ground squirrels. Even

Brown bears will typically nurse their cubs for up to 3 years. The cubs become independent sub-adults shortly after weaning. (NPS/Kelsey Griffin)

of highly nutritional foods, such as those high in fat, protein, and simple carbohydrates (salmon, animal meat, insects, and berries). Their movements track the location of food source seasonality and spatial distribution. By the time brown bears enter their dens in the fall, their body mass is 30%–40% fat; this adaptation is key to their survival through the winter.

Brown bears mate in the late spring and early summer; after fertilization, the embryo grows to only a few dozen cells and does not implant on the wall of the uterus until November or December. If the pregnant mother bear has the fat reserves necessary to support the growth of her unborn offspring, they will continue to grow until they are born in the den in midwinter. The new cubs are helpless to survive on their own (i.e., altricial) as they generally weigh less than a pound (around 0.5 kg) at birth, they lack hair and mobility, and their eyes are closed for the first several weeks of life. Like humans, they require the assistance of their mothers to survive. By the time they emerge from the den, they weigh roughly 10–15 pounds (4–6 kg). All their growth in the den relies on their mother's milk, and they can continue to nurse until 3 years of age. Females can produce 1–4 cubs per litter and must rely solely on stored bodily energy they acquired the previous year to sustain themselves and their litter inside the den. To meet the demands of

with reduced energy demands, bears need to go into hibernation with enough resources to survive the winter. Bears do not eat during the entire time they are in their dens. Each year, they must consume enough calories and nutrients during the spring, summer, and fall to support themselves and their cubs through winter. Typical of carnivores, bears have a relatively simple digestive system. Unlike moose, deer, or cattle, bears cannot efficiently digest and use the nutrients bound in the complex carbohydrates in some foods, such as grasses and twigs. Bear behavior is largely driven by the pursuit hibernation and cub production and rearing, brown bears often lose 40% of their body weight over the course of the hibernation period; this loss is primarily composed of stored body fat. In coastal ecosystems, cubs often stay with their mother for up to 3 years; in Arctic ecosystems, the period of cub rearing may last up to 5 years. As the cubs mature, they transition to a diet of the same food as their mother. Because offspring may continue to nurse for several years, the energy demands on the mother can span the duration of the rearing period. Female brown bears can have their first litter as early as 5 years of age but may not reproduce until they are 7 or 8 years old, especially in ecosystems that are limited in food availability. This late age of first reproduction, relatively small litter sizes, and the number of years the cubs stay with their mother means that each cub is a huge investment for the mother. The survival of each of the cubs produced during her life span (up to 30 years in the wild) is what carries her lineage forward in the population. This investment in offspring is why bears will fiercely defend their cubs when they feel threatened.

SEASONALITY, HOME RANGE, AND HABITAT USE

An important and long-standing area of brown bear research has focused on measuring how bears use the landscape and the habitats that individuals, and thus populations, need to persist at a sustained level. In general, brown bears evolved in open Arctic and alpine tundra and plains habitats, whereas the generally smaller black bears were creatures of the forests across the continent. While these preferences persist, the 2 species do overlap across much of Alaska, and the presence of 1 species can influence how the other uses the landscape. We use a variety of approaches to better understand how brown bears use and select habitats. These include estimating home range sizes (how big an area do bears use) and resource selection (do bears prefer certain habitat types). Together, these analyses can reveal both the total area a bear uses in a year and the critical habitat(s) where it spends most of its time. Finally, related analyses can identify corridors that facilitate the movement of bears from 1 primary habitat to another within or across seasons.

Brown bears have different needs at different times of the year. Because they live in ever-changing seasonal landscapes, they must travel to meet their needs. Outside of the early summer breeding season (often late May to early June) and females with dependent cubs, bears are largely solitary. However, bears do congregate where food resources are clustered, and this is seen most dramatically when bears gather to eat the most valuable foods. This includes, most famously, salmon

Brown bears are selective in choosing denning sites, with the intent of minimizing disturbance while they are hibernating. In the Brooks Range of northern Alaska (pictured here), they choose high-elevation sites protected from the wind where deep snow remains through the winter and acts as an insulative blanket to protect them from the extreme cold of their Arctic environment. (NPS/Mathew Sorum)

and berries but also moths, pine nuts, sedges, and whales washed up on beaches. Despite the availability of high-quality foods, some bears (e.g., females with young cubs) may avoid high-quality areas to minimize the risk associated with interacting with humans and other bears, especially adult males, which are known to kill cubs. Thus, individual bears make decisions by weighing the risk of injury or death to themselves or their cubs versus the reward of high-quality food. Bears also select habitat for reasons other than seeking food, such as temperature regulation, avoiding other predators, or seeking a good den site. Selection of denning habitat has been highly studied, as bears are particularly vulnerable to direct and indirect effects of disturbance during the denning period.

PHYSICAL AND MENTAL PROWESS

Brown bears have long been the source of legend and folklore for their strength and cunning. As authors, we have been fortunate to collectively observe thousands of bears in the wild behaving naturally, responding to our capture attempts, and living in captivity. They are both quick and fast, able to easily outrun any human. They can close short distances in a matter of seconds and easily reach speeds of more than 30 miles per hour (48 km/h) across rugged terrain. Just as impressive, they scale mountains, navigate deep snow, and traverse lakes and rivers. Virtually no topography impedes them. Bears are powerful diggers, with 3- to 4-inch (7- to 10-cm)-long claws that allow them to excavate earth, flip logs, and move aside boulders in search of food or to dig a winter den. Conversely to their tremendous power, their lips and claws are extremely tactile and proficient in consuming small berries or bivalves, and bears can be apparently delicate in the way they handle and position their young cubs when nursing. Their sense of smell is legendary. We have noted bears that "beeline" more than 15 miles (24 km) after picking up the scent of a moose carcass. Their eyesight is believed to be similar to that of humans. Perhaps most surprising, we have also seen bears quickly develop multi-step problem-solving strategies and use tools through trial and error. Through time, we have responded to numerous pleas for help or advice from people challenged by the presence of bears that have entered homes and cabins. Bears survive in a complex and perilous world by relying on their physical strength, dexterity, intellect, and experience. But beyond just survival, these attributes are also on display when we observe bears play. They wrestle as cubs, climb and slide down 200-yard (183-m) snowy slopes repeatedly, chase each other, and appear to be fascinated by novel objects they find in their environment. They truly are a marvel.

VARIATION ACROSS AND WITHIN POPULATIONS

Brown bears are considered a single species across their range due to commonalities in their genetics, morphology, and behavior. However, within the species, there is tremendous variation in the environments in which bears live and how they interact with their environment. One of bears' greatest strengths is their ability to adapt to a wide range of environmental and ecological conditions. Across ecosystems, adult brown bear body size generally

reflects the productivity of the ecosystem they live in and can vary twofold for females and threefold to fourfold for males.[3] As an example, the average adult female on Kodiak weighs approximately 400 pounds (180 kg) in the spring compared to the Arctic, where bears weigh approximately 200 pounds (90 kg).[4] It is not surprising that salmon-rich systems support both larger and more bears (often ≥10 times the population density of salmon-poor populations).[5]

What is most surprising is the level of variation we've seen within a given population. Bears in each population we have recently studied also span a nearly twofold range of adult body sizes. This within-population variation suggests multiple life history strategies and niche use as bears balance risks and rewards.[6] This is further supported by recent work that categorized brown bears based on their dietary habits (i.e., generalist versus specialist). Across ecosystems, populations are composed of individual bears that use a variety of dietary strategies.[7] Bears also vary their behavior at specific life stages as they manage risk (for example, juveniles; females with young, vulnerable cubs), further exemplifying the variability individuals are capable of (i.e., plasticity) in foraging behavior and resulting body size seen across individuals, populations, and the species. In other words, brown bears are individuals, and these individuals find their own, often unique ways to survive and thrive under a wide array of conditions.

THE ECOLOGICAL ROLE
OF BROWN BEARS

Brown bears do not passively exist within an ecosystem; they can play significant roles in nutrient flow within and across terrestrial and marine systems.[8] Because they are omnivores (animals that eat plants and other animals), they directly and indirectly affect a huge array of other species. Important food items for brown bears include ungulates (e.g., moose, caribou, and deer), marine foods (e.g., salmon, flatfish, clams, and beached carrion), insects, and vegetation (e.g., berries and sedges). The importance of these food items varies from ecosystem to ecosystem and from bear to bear. While bears interact with a variety of species, they are most studied as predators of species that are also of high value to human hunters, such as moose and caribou. As a predator, brown bears can directly alter the demographics of their prey populations by killing individuals. As herbivores, they can play an important role as seed dispersers and in stimulating growth of fast-growing species like grasses.

Because bears are critical members of their food webs and require large expanses of intact habitat, the health of their populations can serve as an indicator of overall ecosystem health. For that reason, bears have served as "indicator" or "umbrella" species. Therefore, if there is enough habitat to support a robust population of bears, the habitat should also be sufficient to support many other species with less expansive needs. Similarly, bears can be used to identify sources of stress or threats to an ecosystem.

Many factors threaten the health of brown bear populations. Because of the wide variety of food items they consume, both living and dead (like a wolf-killed moose carcass), bears are exposed to a variety of pathogens. Assessing the presence of pathogens or antibodies can identify potential risks to individual bears as well as document the occurrence of or change in the pathogens present in an ecosystem. Similarly, contaminants and toxins often accumulate as they move up the food chain, and bears can be a good indicator of high levels of heavy metals (natural or anthropogenic), toxins, and micro-plastics. In many areas, bears serve as a sentinel because screens for pathogens, toxins, contaminants, and stress hormones can be conducted in conjunction with other ecological studies simply by collecting and analyzing blood samples.

RESEARCH AND MANAGEMENT
IN NATIONAL PARKS

Ideally, brown bear management in all parks would be guided by a comprehensive knowledge of population trend and size as well as a detailed understanding of their nutritional ecology, habitat needs, and primary threats. In practical terms, this goal is never achieved because resources are limited and Alaska parks are incredibly vast. Rather, biologists use their general understanding of brown bears across their range, in conjunction with local research and monitoring programs, to address a given park's most pressing management issues. These issues vary from one park to the next and within a park over time. Some of these issues are explored further in the individual park chapters in this book. Further, bears live in complex, dynamic ecosystems in a rapidly changing climate. Thus, it is important to replicate work through time to understand the direction and magnitude of change and what this change may mean for brown bears specifically and the ecosystems they reside in generally. To meet this charge, we are constantly striving to refine the questions we ask and develop and improve the methods and tools we use to answer them. While we and our fellow biologists have learned a lot about the ecology of brown bears, there is still much more to learn. With each passing study, the uniqueness of brown bears continues to inspire awe and admiration.

Biologists use their understanding of bears across their range, in conjunction with regular monitoring, to identify any changes that might impact their conservation. (NPS)

NOTES

1. Servheen, Herrero, and Peyton, *Bears—Status Survey and Conservation Action Plan*.
2. Bled et al., "Using Multiple Data Types and Integrated Population Models to Improve Our Knowledge of Apex Predator Population Dynamics."
3. Hilderbrand et al., "Body Size and Lean Mass of Brown Bears."
4. Hilderbrand et al., "Brown Bear (*Ursus arctos*) Body Size, Condition, and Productivity in the Arctic."
5. Hilderbrand et al., "Importance of Meat."
6. Hilderbrand et al., "Body Size and Lean Mass of Brown Bears."
7. Rogers, "Applications of Stable Isotope Analysis to Advancing the Understanding of Brown Bear Dietary Ecology."
8. Hilderbrand et al., "Role of Brown Bears (*Ursus arctos*) in the Flow of Marine Nitrogen in a Terrestrial Ecosystem": Levi et al., "Community Ecology and Conservation of Bear-Salmon Ecosystems."

Habituation occurs when bears learn to tolerate people through repeated neutral, unharmful exposures. Habituated bears are less likely to be disturbed by people and are less likely to charge or attack them. However, bears that become habituated to human foods (food-conditioned) can be very dangerous. (Stuart Leidner)

4 HUMAN–BEAR INTERACTIONS

THE CHALLENGE OF COEXISTENCE

Lindsey Mangipane, Kelsey Griffin, Kyle Joly, Tania Lewis, Buck Mangipane, Patricia Owen, Andee Sears, and Grant V. Hilderbrand

Brown bear behavior is shaped by the personality and experiences of each individual bear. Factors such as food abundance, the density of nearby bears, exposure to humans and human food, age, health, and hunting pressure can all contribute to a bear's behavior when it encounters a person. No one can predict exactly how a bear will behave in any given situation, but there are some broad behavioral patterns that are helpful to be aware of.

Bears eat an incredibly diverse array of food ranging from leaves, roots, and berries of various plants to flesh of mammals, fish, and invertebrates (shellfish or grubs, for example). The diversity and abundance of bear food resources vary across Alaska. In general, foods are more abundant in coastal areas of southern Alaska where salmon are seasonally plentiful and inter-tidal animals are available during every low tide. For this reason, coastal brown bears generally grow bigger and are found in higher densities because the habitat can support more bears than can interior Alaska, where food sources are fewer and less concentrated. When feeding on an abundant food resource like spawning salmon, coastal brown bears can be found in

https://doi.org/10.5876/9781646427116.c004

Food-Conditioned versus Habituated

How a bear interacts with people is based largely on its past experiences. Bears exhibit a high degree of learned behavior; when they are repeatedly exposed to people or are able to receive human-associated food with few consequences, they can begin to behave differently than bears with less experience around people.

Two commonly used terms to describe bears that are involved in conflict situations are *habituated* and *food-conditioned*.

> A *habituated bear* is an animal whose outward response to people has waned as a result of repeated exposure to humans without significant consequence.[1]

> A *food-conditioned bear* is defined as an animal that has learned to associate people, human activities, human-use areas, or food storage receptacles with anthropogenic food as a result of repeatedly accessing anthropogenic foods without substantial consequence.[2]

high densities and appear to tolerate each other in close proximity. We believe bears generalize this tolerance to people, and coastal brown bears have been found to have a much smaller distance at which they react to people than do interior brown bears. This means that coastal brown bears will likely only charge people if surprised at a very close range (less than 30 feet [10 m]). In contrast, interior brown bears often react at far greater distances (over 150 feet [50 m]).[3] Again, each bear is different and every human-bear interaction is unique, but in general, well-fed coastal bears are more tolerant and less likely to respond defensively when surprised.

One important factor that affects a bear's behavior with people is its previous experience with and exposure to people. Bears learn to tolerate people through a process called *habituation*, in which bears consistently encounter people as neutral—not aggressive or threatening or a source of food. Consistent benign encounters teach a bear to fear people less and, therefore, to be less likely to react defensively. In some areas of the state, bear viewing contributes to habituation, which can be an advantage to both bears and people in the right settings. Habituated bears are less likely to be disturbed or displaced by people and less likely to charge or attack humans, something that can benefit both species. However, habituation is not always a good thing. Habituated bear behavior can encourage closer distances between bears and people and may cause bears to be more likely to enter developed human areas, increasing the potential for a bad encounter. Bears that are accustomed to close interactions with people can also be more vulnerable to harvest.

Overall, we have found that these major differences in bear behavior—caused by variation in food abundance, bear density, and exposure to humans and hunting—have led to substantially different bear-viewing rules and recommendations, minimum approach distances, and bear-safety advice across parks in Alaska.

VIEWING BEARS IN ALASKA PARKS

The opportunity to see wildlife is among the primary reasons people visit Alaska's national parks. Brown bears are uniquely charismatic, and many visitors travel to Alaska specifically to have the chance to see them in the wild. Because viewing conditions and settings differ across parks, each has somewhat different best practices for responsible viewing.

Visitors can see bears opportunistically in parks or as part of dedicated bear-viewing trips. Bear viewing is conducted in different ways across Alaska parks (e.g., in boats, buses, or on foot), providing unique viewing experiences (table 1). Most designated bear-viewing sites have high-quality, abundant food resources that reliably attract high densities of bears to the area, such as waters with abundant seasonal salmon returns or salt marshes

with nutritious vegetation. Visitors access these sites by boat or aircraft using authorized commercial operators. Some visitors stay in lodges or camps within the park, while others spend as little as a few hours. People are encouraged to view park websites that provide specific site-based information and regulations before they visit.

Katmai National Park and Preserve has popular interior and coastal locations for bear viewing. The best-known of these viewing areas is Brooks Camp,

The opportunity to see bears is one of the main reasons people visit Alaska's national parks, especially here in Katmai National Park and Preserve. (NPS/Russ Taylor)

TABLE 1. Summary of bear-viewing experience, distance regulations, and bear management practices across Alaska's national parks, preserves, and monuments.

Park	Primary Viewing Experience	Distance Regulations	Recommended Best Practices
Katmai National Park and Preserve	Three viewing platforms at Brooks Camp and fly-in bear viewing in sedge meadows on coast	Approaching bears or occupying space near bears' food source (including fishing) within 50 yards prohibited	Respect bears' space, stay together, use same sites, minimize disturbance, make human use predictable, avoid displacing or food conditioning bears, discourage curious bears from approaching, use trained guides
Lake Clark National Park and Preserve	Fly-in bear viewing in 2 sedge meadows on the coast, boat-based viewing at Crescent Lake	None	Make human use predictable, minimize disturbance, respect bears' space, do not displace bears from feeding, do not let bears access human food, stay in a group, understand bears' behavior, discourage curious bears from approaching
Denali National Park and Preserve	Tour bus or backpacking	Approaching bear within 300 yards prohibited if not in hard-sided vehicle	On tour bus: quiet at wildlife stops, keep all parts of body inside bus, gently nudge animals when bus needs to move Orientation for backcountry users
Glacier Bay National Park and Preserve	Boat-based viewing (both motor vessels and kayaks)	None. Recommend keeping distance of 100 yards from brown bears in vessels to minimize disturbance	Minimize disturbance—if you are changing animals' behavior, you are too close Orientation required for all backcountry users and vessel operators
Northern Alaska National Parks, Preserves and Monuments[a]	Incidental during hiking, rafting, or flightseeing activities	No distance requirement, however, recommended to keep a distance of ¼ mile	Airplanes should stay 2,000' above sensitive areas and harassing animals from airplane is prohibited

a. The Northern Alaska National Parks, Preserves and Monuments category includes Gates of the Arctic National Park and Preserve, Yukon-Charley Rivers National Preserve, Noatak National Preserve, Cape Krusenstern National Monument, Kobuk Valley National Park, and Bering Land Bridge National Preserve.

where bears congregate seasonally to feed on salmon at Brooks Falls. Viewing at Brooks Camp occurs from 4 viewing platforms situated along the Brooks River. Visitors to Brooks Camp must attend an orientation and safety presentation that covers park regulations and safe bear-viewing practices. Access and visitor limits are in place for the viewing platforms. Other interior viewing locations are Moraine Creek and Funnel Creek, where bears fish for salmon during summer months. Viewing at these locations is conducted by hiking and observing bears fishing along both creeks. On the Katmai coast, Hallo Bay and Swikshak Lagoon provide excellent viewing opportunities to see bears foraging in the salt marshes. Geographic Harbor provides visitors with the opportunity to see bears fishing for salmon on the coast from a designated viewing area.

Lake Clark National Park and Preserve has 3 coastal locations where bear viewing is the primary visitor activity: Chinitna Bay, Silver Salmon Creek, and Crescent Lake. Chinitna Bay is salt-marsh habitat with 3 designated viewing areas limited to 20 individuals at one time, with no time limit. Silver Salmon Creek features both salt-marsh habitat and a creek that supports salmon in the late summer and early fall. Visitors at Silver Salmon Creek are not restricted to viewing areas and can move freely and observe bears from various locations throughout the area. Multiple species of salmon return to Crescent Lake, which provides opportunities for

viewing from boats while bears travel and fish along the shoreline and the lake outlet.

Lake Clark and Katmai are the only parks in Alaska that have designated bear-viewing locations; visitors to other parks experience bears as watchable wildlife. In Denali National Park and Preserve, most visitors view bears from the road system, either by bus or, to a lesser extent, personal vehicles. Visitors to Denali also view bears incidentally while engaged in other recreational activities, such as hiking, camping, and biking. Bear viewing in Glacier Bay National Park and Preserve is typically boat-based and can include a multitude of vessels ranging from personal kayaks to cruise ships.

Bear viewing in Lake Clark includes beach and salt-marsh habitats. Visitors can move freely and observe bears from various locations. (NPS/Kevyn Jalone)

Visitation to bear-viewing locations is reported only in Lake Clark and Katmai National Parks and Preserves. In 2019, the last year for which complete, non-pandemic-influenced data are available, 10,711 and 7,917 visitor-use days were reported for bear viewing in Katmai and Lake Clark, respectively.[4]

Since the early 2000s, bear-viewing visitation has been increasing. While the numbers may appear rather low, much of the visitation is concentrated at a limited number of small sites over a short period of time. With increases in visitation likely to continue, park managers will be challenged with determining what level of visitation negatively impacts visitor experience, bear behavior, and bear populations.

Overview of Bear-Viewing Best Practices

Whenever interactions between bears and people occur, there is always the potential for negative impacts to both bears (e.g., disturbance) and humans (e.g., safety). However, by following best practices, people can greatly reduce these risks and successfully view bears with minimal impact. Each park is unique; as discussed above, various factors influence bear behavior and the resultant best practices for each area. We encourage visitors to view park-specific information before their trip to learn the most current guidance for staying safe and reducing impact to bears.

Consistent human behavior is one of the primary tenets for safely viewing bears while minimizing disturbance.[5] This behavior can include regularly using the same viewing sites and accessing those sites in a consistent manner. Predictable use of the landscape by humans can reduce the potential for surprise encounters with bears and minimize the level of disturbance because bears are less likely to be displaced from an area when they can predict patterns of human activity.[6]

In addition to using the same areas consistently, appropriate and predictable human responses to bears can minimize the potential for negative interactions between bears and people. Bears should be given multiple travel paths that allow them to leave the area where viewing is occurring. They should never be approached, crowded, or otherwise pursued. Some parks have specific regulations on the minimum distance visitors can be from bears (table 1). However, the suite of factors that influence how visitors and bears are managed differs across parks; there is no universal minimum approach distance. In addition, some parks have no specified regulation for how close people can be to bears, recognizing that individual bears may have different tolerance levels. In these parks, visitors are asked to respect bears' "personal space" and refrain from being close enough to alter their behavior. Just as people should not approach bears, bears should not be able to approach people. Young bears, in particular, will sometimes test boundaries and may approach people out of curiosity or, in more serious situations, to obtain human food. If a bear continues to approach after visitors ensure that they are not blocking its travel route, visitors should assert themselves to defend their personal space, starting at a low intensity level (such as holding their ground and talking to the bear) and escalating their response as appropriate. Groups of people are less likely to be attacked by a bear than are individuals.[7] Therefore, it is recommended that visitors remain in groups when in bear country and especially while viewing bears. In addition, some parks recommend that visitors are accompanied by a trained guide who is well versed in bear behavior and safety (table 1). Bear safety and interaction guidelines are discussed in more detail later in this chapter.

Potential Impacts of Bear Viewing on Bear Behavior

Best management practices can reduce impacts to bears; however, human activity can still elicit a suite of behavioral responses from bears, ranging from increased vigilance to shifts in how they access areas in space, time, or displacement. Even if bears do not display an overt behavioral response to disturbance, they may still experience a physiological response such as increased heart rate or the release of stress hormones. However, brown bears exhibit a high level of among-individual variation in behavior, and responses to disturbance caused by bear viewing can be highly variable across individuals or populations. Factors such as the timing, frequency, and duration of disturbance; bear density; sex and age of bears; and resource availability can all influence an individual's response. In areas such as Yellowstone National Park, many bears have learned to tolerate the presence of large numbers of people at roadside "bear jams" in exchange for access to high-quality natural food sources that are present along roadsides.[8] Similar patterns are observed among bears at Brooks Falls in Katmai National Park and Preserve, where bears tolerate high levels of human visitation as they congregate to feed on spawning salmon.

Although bears in some locations appear to be resilient to a certain level of human-caused disturbance, in other areas bear viewing causes individuals to significantly alter their behavior, resulting in reduced feeding opportunities and increased energetic costs.[9] Some male bears have been found to forage at night, shifting activity to times when humans are not present.[10] Because large-bodied males have high energetic requirements, constraining the amount of time spent foraging could impact their ability to meet their nutritional needs.[11] Although the presence of humans can deter some bears from using areas where viewing occurs, others have learned to use bear-viewing areas as a refuge to avoid individuals that are more dominant.[12] For example, females with young cubs will sometimes use areas when bear viewers are present to avoid dominant males that could otherwise prevent them from accessing preferred food resources or present a risk to their cubs.

Increased exposure to people through bear viewing can also result in bears losing their fear of people and ultimately becoming habituated to human activity. Habituation can be beneficial in reducing a bear's likelihood of responding defensively to people. However, habituation to humans can also have direct negative impacts on both bears and people and can result in increased human-bear conflicts as well as a higher risk of mortality through vehicle collisions, harvest, defense-of-life kills, or management actions.[13] For example, if a bear is outwardly tolerant of people, it may embolden humans to engage in bad behavior such as approaching bears too closely to improve viewing or photo opportunities. Close approaches may increase bears' stress levels or potentially cause a bear to respond defensively, increasing the risk of human injury or death. Ultimately, these types of negative interactions often lead to the death of the bear as well. Bears that lose their fear of humans may be more likely to spend time near populated areas where human-associated food sources are more readily available. Over time, receiving food rewards can lead to bears associating people with

food (otherwise known as food conditioning), which can present a considerable public safety concern and often results in bear mortality. In addition, habituated bears may be more vulnerable to harvest, resulting in an ethical dilemma in terms of managing bear viewing in areas where harvest occurs.

Conservation Benefits of Viewing Programs

Although bear viewing can negatively impact bears, we believe that viewing provides a multitude of conservation benefits. Seeing a bear in the wild can be a transformative experience for a person. Their size, intelligence, resourcefulness, playfulness, and famously protective nature of their cubs help people easily connect with brown bears. These connections are often the source of motivation for people to actively engage in activities that directly and indirectly benefit brown bear conservation. Seeing bears do interesting things in the wild often motivates people to want to learn why they do them and more about their ecology. In many places, like Brooks Camp, viewing can be coupled with educational outreach programs, such as ranger talks. This learning helps people understand what bears need and can lead to lifelong support of brown bear conservation efforts. People who have never seen a bear before can go from awestruck to knowledgeable to advocate in a single bear-viewing trip.

Bears are also an extremely charismatic species and often act as ambassadors for the systems they live in. A prime example is "Fat Bear Week" in Katmai National Park and Preserve. In recent years, the fat bears that gather at Brooks Falls have gained national and international attention. Each fall, people from around the world vote to determine which bear is the fattest. In 2024, more than 1 million votes were cast by people from over 100 countries. The considerable attention this event has brought to the region has also served as a way to highlight important conservation issues, such as the proposed Pebble Mine, which has been a significant conservation concern in the region for more than a decade. Proposed to be the world's largest open-pit mine, Pebble Mine could directly affect these bears, as well as the entire Bristol Bay watershed. Although most people will never travel to Alaska, let alone to the Bristol Bay region, Fat Bear Week has given many people a reason to care about the area and the resources it supports.

Fat Bear Week is a much-anticipated competition that attracts hundreds of thousands of voters throughout the world. Bear #151 (Walker) is a frequent competitor for the fat bear title, but has yet to win. (NPS/Noami Boak)

Bear #128 (Grazer) was the 2023 Fat Bear champion. (NPS/J. Coumo)

TOOLS AND ADVICE FOR HUMAN-BEAR INTERACTIONS

National parks were established to protect and provide opportunities to visit our most amazing natural places. While unique natural features and scenic beauty are characteristics common to these places, many were also established for their importance as habitat for the wildlife native to North America. As people began to visit parks, interactions with wildlife inevitably occurred. In the early years of parks, bears opportunistically exploited human food and trash that were the by-products of increased visitation. This food conditioning led to increases in human-bear conflicts, resulting in human injuries, property damage, and an increase in the number of bears killed by park managers. Early management efforts focused on removing bears engaged in conflict, but this did little to reduce human injuries and property damage. This informal approach to management was ineffective, and concerns stemming from this situation led parks to develop a more proactive bear management approach focused on both human and bear behavior. This approach combined visitor education about bear behavior, proper food and waste management, development and use of bear-resistant garbage cans, and removal of bears that damaged property or were potentially hazardous to visitors. With improved education and increased visitor compliance with proper food and waste management requirements, human-bear conflicts were dramatically reduced.

Brown bear attacks have never been common in national parks or elsewhere, and the numbers have been greatly reduced through bear-safety education and management efforts. For example, despite relatively high visitation, Denali National Park and Preserve has only recorded 27 human injuries and 1 death related to human-bear conflicts since 1946. Parks with lower visitation rates have had even fewer human injuries or deaths related to conflicts, with Lake Clark recording 2 human injuries and Gates of the Arctic reporting 4 injuries and 1 death since 1990. Although some parks have detailed human-bear interaction data, many do not. Efforts have been made to standardize data collection across Alaska parks; however, work is still needed to ensure accurate reporting.

Preventing bear-human conflicts starts with education, and education starts with basic safety precautions and best practices for recreating in bear country. Being "bear aware" is an important step for preventing a bear encounter. This entails being observant for bears and the signs their presence leaves on the landscape, such as scat and tracks. This awareness combined with general safety tips like staying alert, being visible, making noise, traveling in groups, never leaving gear unattended, and storing food and gear properly help minimize the risk of ending up in a negative encounter. Follow-

Whenever bears and people interact, there is potential for either or both to be negatively impacted. Those negative impacts can be mitigated with best practices. It is always a good idea to give bears space. (NPS/Rebekah Jones)

ing best practices increases human safety and also helps keep bears wild.

When camping in bear country, we highly recommend the use of an electric fence and bear-resistant food containers (BRFC); in many parks, they are required. The idea is to separate yourself from any food attractants and keep enticing food and smells from bears. Using BRFCs ensures that all food and other potential attractants (e.g., toothpaste and other toiletries) are not accessible to bears. Cookware should be cleaned well off main trails and below high tide lines in coastal areas. Food should be prepared and consumed at least 100 yards (100 m) from your tent site and food storage area—downwind of your campsite, if possible. Cooking-area selection should prioritize good visibility, when possible, to help minimize the risk of a surprise encounter with a bear passing through the area. In a coastal park, food preparation and eating should be done in the inter-tidal zone when possible. This keeps cooking smells away from camp and storage areas and will allow the next tide to wash away any food particles. At night, the BRFC and any other items that might have a food scent should be at least 100 yards (100 m) from your tent and hidden in vegetation or rocks when possible.

When recreating in bear country, you should prepare for the possibility of a bear encounter. There are a number of reputable resources for learning how to respond to a bear encounter. The book *Bear Attacks: Their Causes and Avoidance* is a valuable resource that has been referenced in National Park Service (NPS) safety guidance. The NPS has also developed number of bear-safety publications such as *Bear Safety in Alaska's National Parklands*, developed specifically for Alaska parks.[14]

We strongly recommended carrying a deterrent and having it in an easily accessible location. It is advisable to check with parks for specific recommendations that work well in the location and conditions you are planning to visit. Bear spray is highly recommended due to its proven efficacy under a variety of conditions.[15] In a 2008 study, bear spray was found to stop a brown bear's undesirable behavior 92% of the time,[16] preventing all but a small number of minor injuries to people that did not require hospitalization. Although many voice concern about the potential for wind to negatively affect the performance of bear spray, it was only found to influence the outcome in 7% of incidents.[17] Under laboratory conditions, bear spray has been found to hit a target that is 6 feet (2 m) directly in front of the user even under high (>20 mph [10 m/sec]) headwind and crosswind scenarios.[18] For bear spray to be most successful, carry it in a location where it can be accessed quickly in a surprise encounter, such as in a hip or chest holster, and practice using it in an appropriate place before you need to use it.

When you encounter a bear, it should trigger a series of actions you follow to decrease the likelihood

of the encounter becoming a negative human-bear interaction. It all begins with you observing a bear, so keep your eyes scanning at all times while moving through bear country. If the bear does not appear to see you, move away from it to limit the chance of an encounter. If the bear is aware of you, assess its behavior (as described below) to evaluate the type of encounter. Encounters can be non-defensive or defensive, and the actions taken will vary depending on the type of encounter. If the bear is aware of you but shows no interest, is traveling steadily, or engages in a behavior such as feeding, this can be considered a non-defensive behavior. In this situation, you should change your course of travel to increase your distance from the bear. Remain observant, watching for any behavioral changes the bear may make.

Bears are often attracted to anything people intentionally or accidentally leave behind. (NPS/Matt Harrington)

Bears are naturally curious and will investigate anything left unattended. (NPS/Kelsey Griffin)

continues to approach, elevate your actions. This could include raising your voice or shouting. If the bear departs at this point, the encounter has ended. If it continues to approach, stand your ground and become more assertive by escalating your shouting or using a noisemaker, such as an air horn, if available. During this period, prepare a deterrent. If the bear charges, hold your ground and remain assertive; if it is within range of your deterrent, use it. If the bear makes contact, fight back vigorously, as this is likely a predatory attack. Kick, punch, or hit the bear's face, eyes, and nose.

Defensive encounters may arise when bears are defending food or a female bear is defending her cubs. Defensive encounters are typically sudden and at close range. Behaviors indicative of a defensive bear may include snorting, huffing, jaw popping, and charging. Any of these behaviors should elicit an immediate response from you, such as stopping and standing your ground. In a defensive situation, the goal is to remove yourself as a threat to the bear. Talk calmly to the bear and move slowly away if it is stationary. Continue to monitor the bear as you move from the area. If the bear advances, stop and stand your ground. If the bear charges, remain non-threatening and stand your ground, as most charges do not end in contact. If you have bear spray, use it. If the bear makes contact in a defensive encounter and it is a brown bear, play dead. Lie face-down, hands clasped behind your neck, with your

If the bear approaches you, ensure that it is aware of your presence. You can do this by talking calmly to the bear and standing your ground. At this point, your actions should be calm and of low intensity; however, if the bear is making a direct approach, you should prepare your deterrent. The actions you take in a human-bear interaction should follow a continuum of increasing intensity that corresponds to the bear's behavior toward you. If you are in a group, remain together. If the bear is focused on you and legs spread apart so the bear has a harder time turning you over. Do not move until the bear completely leaves the area. If the attack is prolonged and the brown bear begins to feed on you, fight back vigorously, as the encounter has changed from defensive to predatory. If it is a black or polar bear, do not play dead and fight back vigorously, as most black and polar bear attacks are predatory. Fight back against any bear that attempts to enter your tent.

If you are involved in a human-bear interaction, make sure to report it to the park in which it occurred once you are safe. These data are important for managers so they can assess if the incident requires management action, such as posting public notices or closing areas. Human-bear interaction data are also important for learning more about how to prevent and respond to negative interactions between bears and people. These data can ultimately help managers identify problem areas, assess the efficacy of bear-safety measures, and inform future management.

IMPACTS OF DEVELOPMENT ON BEARS

The westward expansion of settlers across North America greatly reduced the geographic range and abundance of brown bears due to a combination of factors, including fear and intolerance of predators, protection of livestock, bounty hunting, habitat loss or exclusion, and population fragmentation. As a result of Alaska's remoteness, relatively sparse human population, and limited development, brown bears here have never been considered threatened or endangered, unlike the situation in the Lower 48 states. Since the 1960s and 1970s, much of the focus of conservation and recovery efforts for brown bears in the Lower 48 states has been on minimizing mortality and maintaining and enhancing connectivity between ecosystems and populations. Significant research has informed the management of protected and small brown bear populations in the Lower 48 states and western Canada and has noted the varied, direct, and indirect adverse effects associated with development.

Roads, in particular, have been shown to have negative impacts on individual brown bears by restricting travel, increasing energetic costs, and limiting access to key resources.[19] Both black and brown bears have been shown to avoid areas close to roads and trails,[20] effectively reducing the habitat available to them and changing their natural behaviors.[21] In addition to these indirect impacts, mortality rates for bears tend to be higher as proximity to roads increases.[22] While vehicle strikes do contribute to mortality, management actions taken to remove bears that come into conflict with people are often more numerous. For example, the most significant factors for bear mortality in Banff and Yolo National Parks in western Canada were trail development, campgrounds, human food, and garbage. Roads can also facilitate additional (legal and illegal) harvest of brown bears.[23] Thus, it is often the access and infrastructure that accompanies road development that results in the majority of individual bear deaths and issues.[24]

At the population level, statistical modeling efforts reveal that road densities above certain thresholds result in negative impacts on population trends, largely due to reduced survival of sub-adult bears and females with young cubs.[25] Despite the impacts of roads and motorized access on habitat use, home range selection, movements, population fragmentation, survival, reproductive rates, and population density, motorized access management—particularly of industrial roads—can mitigate these adverse impacts.[26]

Though parklands in Alaska are composed mostly of large, undeveloped, and protected expanses of land, some park infrastructure can be substantial (e.g., the Denali National Park and Preserve visitor center and headquarters complex) and include access roads (e.g., the Denali Park Road and the McCarthy Road in Wrangell–St. Elias National Park and Preserve). Park infrastructure may also be centered in habitats that are highly attractive to bears (e.g., Brooks Camp). Existing inholdings (private lands embedded within park boundaries) occur, and are actively accessed and used, in most Alaska parks for such legal activities as mining, tourism, and recreation. In the future, 2 large ore deposits present a particular suite of challenges to Alaska park managers. In Gates of the Arctic National Park and Preserve, an industrial access road through the Kobuk River portion of the preserve was proposed to allow transport of supplies and products to and from the Dalton Highway and the Ambler Mining District (see chapter 9). The proposed Pebble Mine site lies to the south and west of Lake Clark National Park and Preserve and is a long-standing and controversial development project at the headwaters of the Bristol Bay fishery (see chapter 11). Any direct impacts to the fishery have the potential to adversely affect bears in the park and the entire ecosystem. In addition, the access road, which is proposed to come within 6 miles (10 km) of the park, as well as the influx of people living and working onsite carry the potential for the direct and indirect impacts discussed above. Thus, an understanding of the direct and potential indirect effects of development and infrastructure, both existing and proposed, is crucial for management of NPS lands in Alaska as we strive to maintain natural processes and behaviors of wildlife generally and bears specifically.

BEAR HUNTING

In the Alaska National Interest Lands Conservation Act (ANILCA) of 1980, which created or expanded most of Alaska's parklands, the United States Congress acknowledged that humans are a component of natural ecosystems and, thus, hunting is an important aspect of wildlife conservation and management in Alaska. In contrast to most parks across the United States, more than 80% (~45 million acres) of Alaska parklands are open to harvest; the bulk of this acreage includes almost the entirety of all the new areas created by ANILCA. Unlike the Lower 48 states, brown bears have never been designated as threatened or endangered in Alaska, so the harvest of brown bears is allowed across most of the species' range in the state.

Alaska is home to an estimated 30,000 brown bears, and hunters kill an average of 1,500 to 2,000 bears each year.[27] About two-thirds of these brown bears are harvested by individuals who are not Alaska residents. While subsistence hunting is allowed on most parklands, the vast majority of this harvest occurs outside of parks.

The State of Alaska manages most of the harvest in the state (but not in national parks or "national park monuments" described in ANILCA), and it governs using a sustained yield approach that is achieved through numeric goals for both population size and harvest level (Alaska Constitution, Article VIII, §4; Alaska Statutes §16.05.255(k)). In contrast, the NPS focuses more broadly on ecosystem function and the maintenance of natural processes and behaviors.[28] To this end, predator control (culling) of bears to increase moose, caribou, and other populations of prey species is not allowed on parklands (parks, preserves, or monuments).[29] Similarly,

the hunting of bears with the aid of processed foods as bait is problematic on parklands due to public safety concerns and the increased potential for negative human-bear encounters and for altering natural behavior patterns.[30] Thus, the management goals, allowable activities, and data needs differ between the State of Alaska and the NPS. If state harvest regulations conflict with NPS mandates in national preserves, NPS rules supersede those state regulations as outlined in 36 C.F.R. §13.42(a).[31]

Even more than 40 years after the passage of ANILCA, the task of allowing and managing harvest on parklands while ensuring that wildlife populations are managed for natural processes, abundances, and behaviors remains extremely challenging. Given that subsistence hunting was authorized in almost all ANILCA-created parks and monuments and "sport" hunting in all Alaska preserves, it is apparent that both types of hunting and some level of harvest are compatible with the Organic Act (1916), ANILCA, and NPS management policies. But given the multiple regulatory entities involved (see chapter 1)—each with different missions and objectives—understanding, setting, and observing hunting regulations is complicated for both managers and the public. How much harvest can be allowed before it impacts natural abundances? Do fair-chase principles come into play with harvest for "sport" purposes? And, if so, who defines fair chase? Are harvest practices, like baiting bears

Sustained Yield

The state constitution mandates that "fish, forests, wildlife, grasslands, and all other replenishable resources belonging to the State shall be utilized, developed, and maintained on the *sustained yield principle*, subject to preferences among beneficial uses" (Alaska Constitution, Article VIII, §8, emphasis added). State law defines sustained yield as "the achievement and maintenance in perpetuity of the ability to support a high level of human harvest of game, subject to preferences among beneficial uses, on an annual or periodic basis" (Alaska Statutes §16.05.255(k)(5)). It further provides that "high level of human harvest" means the allocation of a sufficient portion of the harvestable surplus of a game population to achieve a high probability of success for human harvest of the game population based on biological capabilities of the population and considering hunter demand (Alaska Statutes §16.05.255(k)(2)).

with human-placed foods that alter their natural behaviors, consistent with the way parks are to be managed? How does the NPS balance providing an opportunity for harvest with methods that have public safety consequences? These types of questions represent an area of wildlife policy that needs to be much more thoroughly developed to move NPS management of wildlife forward in future generations.

The multi-agency regulatory construct also has implications for how well managers understand the impacts of harvest. For example, hunters are asked to report harvests to either the federal subsistence program or the State of Alaska, depending on the specific hunt. Information requested by the 2 entities is different based on their different needs. For example, hunters reporting a bear harvest are asked to report game management unit, subunit, or drainage, but not if they were in a national preserve (where boundaries overlap). Given that not all harvests are reported and location details for reported harvests may be imprecise, lacking, or inaccurate, determining the actual harvest numbers on parklands can be very difficult. Increasing cooperation among agencies collecting harvest data and wildlife management agencies, like the NPS, as well as improving reporting compliance and accuracy could greatly benefit the management and conservation of wildlife—including brown bears—in Alaska.

Broadly, collaboration across agencies on research and management activities is in the best interest of all managers because, regardless of mandates, decisions by both the State of Alaska and the National Park Service benefit from current and rigorous information.

NOTES

1. Herrero et al., "From the Field"; Hopkins et al., "A Proposed Lexicon."
2. Lackey et al., *Ursus*.
3. Smith, Herrero, and DeBruyn, "Alaskan Brown Bears."
4. National Park Service, "Visitor Use."
5. Penteriani et al., "Consequences of Brown Bear Viewing Tourism."
6. Wilker and Barnes, "Responses of Brown Bears to Human Activities at O'Malley River."
7. Gunther and Haroldson, "Potential for Recreational Restrictions to Reduce Grizzly Bear–Caused Human Injuries."
8. Haroldson and Gunther, "Roadside Bear Viewing Opportunities in Yellowstone National Park."
9. Fortin et al., "Impacts of Human Recreation on Brown Bears (*Ursus arctos*)."
10. Rode, Farley, and Robbins, "Behavioral Responses of Brown Bears."
11. Rode, Farley, and Robbins, "Behavioral Responses of Brown Bears."
12. Nevin and Gilbert, "Perceived Risk, Displacement, and Refuging in Brown Bears."
13. Herrero et al., "From the Field."
14. National Park Service, *Bear Safety in Alaska's National Parklands*.
15. Smith et al., "Efficacy of Bear Deterrent Spray in Alaska"; Smith et al., "An Investigation of Factors Influencing Bear Spray Performance."
16. Smith et al., "Efficacy of Bear Deterrent Spray in Alaska."

17. Smith et al., "Efficacy of Bear Deterrent Spray in Alaska."
18. Smith et al., "An Investigation of Factors Influencing Bear Spray Performance."
19. Mattson, Knight, and Blanchard, "The Effects of Developments and Primary Roads on Grizzly Bear Habitat Use in Yellowstone National Park, Wyoming."
20. Kasworm and Manley, "Road and Trail Influences on Grizzly Bears and Black Bears in Northwest Montana."
21. McLellan and Shackleton, "Grizzly Bears and Resource Extraction Industries."
22. Benn and Herrero, "Grizzly Bear Mortality and Human Access in Banff and Yoho National Parks"; Boulanger and Stenhouse, "The Impact of Roads on the Demography of Grizzly Bears in Alberta"; Proctor et al., "Effects of Roads and Motorized Access on Grizzly Bear Populations in British Columbia and Alberta, Canada."
23. Mace et al., "Relationships among Grizzly Bears, Roads, and Habitat in the Swan Mountains, Montana."
24. Benn and Herrero, "Grizzly Bear Mortality and Human Access in Banff and Yoho National Parks."
25. Boulanger and Stenhouse, "The Impact of Roads on the Demography of Grizzly Bears in Alberta."
26. Proctor et al., "Effects of Roads and Motorized Access on Grizzly Bear Populations in British Columbia and Alberta, Canada"; Lamb et al., "Effects of Habitat Quality and Access Management on the Density of a Recovering Grizzly Bear Population."
27. Alaska Department of Fish and Game, "Brown/Grizzly Bear Hunting in Alaska."
28. National Park Service, *Management Policies 2006*, chapter 4, "Natural Resource Management."
29. National Park Service, *Management Policies 2006*, chapter 4, "Natural Resource Management," §4.4.3.
30. Herrero, *Bear Attacks*; Glitzenstein and Fritschie, "The Forest Service's Bait and Switch."
31. National Park Service, "Hunting and Trapping in National Preserves."

When watching bears, it is important to closely watch their behavior. Experienced guides and prudent practices help keep visitors safe. (Stuart Leidner)

5 CLIMATE CHANGE

CAN BEARS ADAPT TO A DYNAMIC WORLD?

William Deacy, Kyle Joly, Nina Chambers, David D. Gustine, Grant V. Hilderbrand, Kim A. Jochum, Pam Sousanes, and David K. Swanson

Human-caused climate change is the fundamental conservation challenge we face today.[1] In the Arctic, the rate of warming is 4 times faster than the global average.[2] The effects are unmistakable across the landscape: vegetation communities that bears and other wildlife use for habitat are changing due to warming air temperatures, shifting precipitation patterns, and increasing wildfire activity.[3] Salmon, a major food source for some bear populations in Alaska, are impacted by warming ocean temperatures, streamflow, and water temperatures. Brown bears are both directly and indirectly affected by all these changes.

◄ Bears seek shade or cool themselves in water on warm summer days. (NPS/Jim Pfeiffenberger)

BEAR BEHAVIOR

Increasing temperatures will likely directly change bear behavior in a few ways. Bears tend to seek shade, water, or snow patches when they get too hot.[4] In a warmer Alaska, they will likely need to spend more time staying cool, which could decrease the time they have to seek food and shift the locations where they spend time on the landscape. For example, they may spend less time on sunny southern slopes and more

https://doi.org/10.5876/9781646427116.c005

Warming temperatures will alter stream flow dynamics. Bears, like people, have a harder time fishing if water levels are too high. (NPS/Lian Law)

time on shaded northern slopes or in higher elevations as temperatures warm. Warmer temperatures and changes in precipitation are also likely to affect denning.

Denning is an adaptation to living in seasonal environments where cold temperatures and a lack of food make resting in dens a better strategy than staying awake through winter.[5] This is reflected in denning patterns across North America, where bears living in colder northern latitudes stay in dens longer than bears in warmer southern ranges. Bears in Alaska will likely enter dens later in the fall and emerge earlier in the spring in response to changes in food availability and temperature.

Temperatures in Alaska have warmed in all seasons, but winter temperatures have increased the most dramatically.[6] On average, warmer winter and spring temperatures have decreased the snow season by 2 weeks in parks in northern Alaska over the past 20 years.[7] Bears in Alaska could benefit from being active more days per year to gain weight and from a shorter period of denning when they must survive on their stored energy reserves. But shorter denning periods could cause an increase in human-bear conflicts due to the longer time when bears and people could interact, as well as bears being out of their dens when people don't expect them to be.

How snowfall will change with climate warming is not yet well understood. In general, Alaska is projected to become wetter. Greater moisture carried by a warmer atmosphere could result in more snowfall. However, increasing temperatures may turn that precipitation into rain. Denning bears primarily use snow and soil to insulate themselves from cold air. One hazard created by climate change is the increased chance of winter rainfall. Rain has the potential to destabilize earthen dens and melt the snow that insulates them from outside frigid air. This could cause a bear to have to move to a new denning site during winter, which could be very costly for adults and fatal for young cubs.[8] Bears could adapt to these changes by denning at higher elevations and on colder, north-facing slopes. Mid-winter warmups also cause rain-on-snow events that can form thick, persistent, and hard ice layers that are detrimental to many wildlife species.[9]

BEAR FOODS AND FORAGING

Brown bears are generalist omnivores, which means they eat a wide variety of both plant and animal foods. This eating strategy has the advantage of flexibility. Bears are not dependent on a single food and can switch among foods as they become more or less available, taking advantage of seasonal patterns. On the one hand, this puts brown bears in a good position to react to the many expected changes in food availability brought about by climate change. On the other hand, they are ecologically connected to many other species,[10] which means that when climate change directly impacts a food resource (like sedges, salmon, or berries), it will indirectly change where, when, or how much of that food bears can eat.

Phenology, the Timing of Natural Events

Air temperature controls the development and growth rates of plants and animals, which affects phenology, the timing of key life events. Animal migration, berry ripening, and the spring flush of plant growth are examples of events that are sensitive to warming. Changes in phenology are important to animals like brown bears that must time their movements to take advantage of ripening berries, spawning salmon, and the growth of fresh shoots of vegetation. Bears can only eat their most valuable foods if they are in the right place at the right time. Where and when to forage for the best food items is knowledge often transmitted from mothers to cubs. With climate change altering the timing of when these foods are available, brown bears may be challenged with learning new foraging patterns.

Changing precipitation patterns and warming temperatures have led to longer growing seasons. For example, in northern Alaska, the growing season has lengthened by an average of at least 2 weeks over the past 20 years.[11] A longer growing season can lead to increased primary productivity, that is, more plant growth. This could increase the abundance of bear foods such as forbs and grasses as well as berry crops—blueberries, crowberries, and soapberries. The longer growing season has also led to the expansion of shrubs and greening of the Arctic, which benefits moose at the northern reaches of their range—another food source for bears.

Alternatively, hot summers with little precipitation may reduce the berry crop and accelerate plant development. Rather than an extended period of time when plants develop slowly, warming and drought can collapse the growing season into a shorter period.[12] This matters to bears and the many other animals that can only digest the tender, most nutritious parts of plants that are usually only available during the early and middle growing season. Tough, woody, mature plants can only be digested by specialist herbivores, such as moose.

Ungulates Are an Important Food Source for Bears

Bear population density tends to correlate very closely with the amount of meat in bear diets.[13] Most bears live in a place that has abundant meat resources. Therefore, how climate change affects the amount and availability of salmon and ungulates (e.g., moose, deer, and caribou) is particularly important to how brown bear populations in Alaska will respond to climate change. Moose habitat is predicted to expand in Alaska, and there are an increasing number of reports of moose moving into the Arctic as shrubs and trees invade.[14] Moose, especially newborn calves, are an important food source for brown bears. The increase in the availability of moose and expansion of their range could be beneficial to brown bears. In contrast, caribou habitat is predicted to decline with climate change.[15] Declines in caribou populations and changes to migration routes and calving grounds could decrease brown bears' access to caribou calves and carrion.

In addition to ungulates, brown bears prey on other mammals, such as ground squirrels. In general, tundra-dwelling prey species are predicted to have less habitat while boreal forest-dwelling species are predicted to expand.[16] Brown bears compete with other mammals, such as wolves, for food. Therefore, changing distributions of other predators may impact brown bears. One such example is polar bears, which normally depend on sea ice and feed on marine mammals. Due to the loss of sea ice, polar bears spend more time on land and are finding new sources of food. This can lead to competition between the 2 bear species for land-based foods, but, interestingly, it has also led to interbreeding and the birth of hybrid offspring. So, potential genetic modification of the species related to climate change is something to watch.

How Salmon Resources May Change

Pacific salmon, a key food for bears in Alaska, are expected to have a mixed response to warming temperatures. Before entering the ocean, salmon spend time growing in the freshwater streams, rivers, and lakes where their parents spawned. In general, warmer water tends to contain more food for salmon, which is why some populations of sockeye salmon are expected to benefit from climate change. However, excessive warmth can lower the amount of dissolved oxygen in streams to levels too low for fish to survive. For example, in 2014, the Kobuk River, located right at the Arctic Circle, experienced a widespread die-off of chum salmon associated with particularly warm waters.

Where salmon populations increase or expand their range, bears are likely to thrive. However, bears that depend on salmon spawning in warmer waters might experience smaller runs as temperatures increase. In addition to uncertainty about future salmon populations, it is also unclear how increased stream and river flows from increasing precipitation and glacier melt will alter bears' ability to catch salmon, even if they were more abundant. We have seen that bears, like people, have a more difficult time fishing in most places when water levels are very high. Because of the importance of salmon to brown bears in Alaska and the degree of uncertainty as to how climate change will impact salmon, this is one of the most important issues for bear researchers, biologists, and managers to get a handle on for future conservation of the species.

Ungulates are an important food source for many bears. As moose habitat expands into the north, moose become an even more important source of meat for bears in that region. (NPS/Ken Conger)

Pacific salmon have shown a mixed response to climate change. Salmon runs have become less predictable, and some shallow streams have experienced die-offs as water temperatures have become too warm. Changing salmon availability will undoubtably impact bears. (Mark Johnson)

WILDFIRE IMPACTS ON HABITAT

Wildfires are one of the most important natural drivers of ecological change in the boreal forest, where many species, like black spruce, are adapted to them. Wildfires reset old growth forests to allow younger (seral) stage communities to colonize, such as grasses and then shrubs. Deciduous trees, such as birch, tend to take over after the early colonizers and then, over time, spruce return to dominance. However, climate change increases the frequency and severity of fires, which could lead to a shift from coniferous species like spruce to more deciduous forests of birch and aspen.[17] Drought conditions may also lead to fires in areas that don't typically burn and have a longer period of recovery.

Increasing fires alter habitat of brown bears and other boreal species. In the short term, fires destroy plants that bears and ungulates eat. Bears tend to find more food in the early succession plant communities (forbs/grasses) that establish in the year or two following a fire compared to the large, mature tree communities that tend to establish after longer periods without fire. Fire severity, how hot a fire burns, will impact which plants return after a burn. This could be important for brown bears, depending on what plants return and what plants do not. In general, fires increase land-cover diversity, which tends to give omnivores like bears a greater variety of foods to eat.

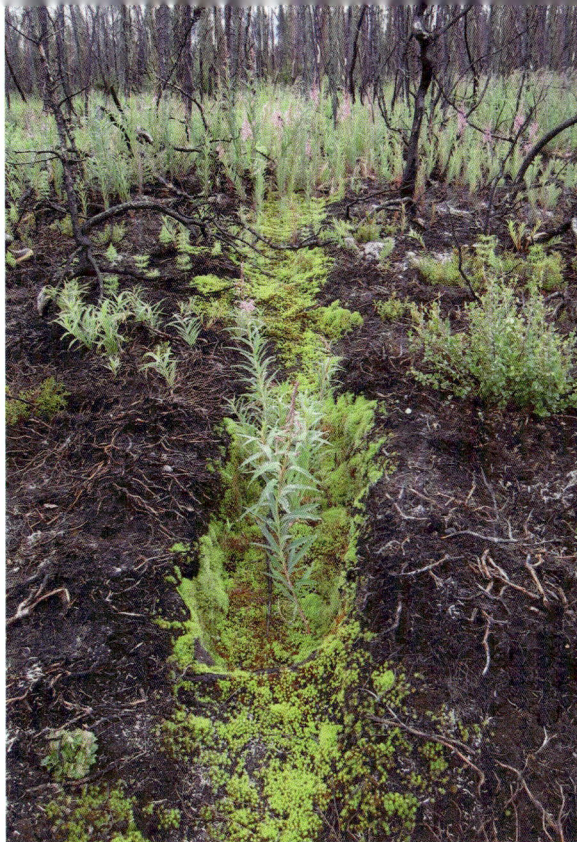

Wildfires are increasing in occurrence and severity. How hot a fire burns impacts what plants return and generally increases diversity, which can increase the kinds of food available to bears. Here we see how the depressions from bear tracks through a burn are helping reestablish vegetation. (NPS/Amy Miller)

CLIMATE CHANGE, BEARS, AND PEOPLE

Ecosystems are highly connected networks, so environmental change caused by climate warming touches every aspect of them. This includes people, who both strongly affect and depend on nature. We cannot view the environment and our ecosystems as separate from people. Humans impact natural resources across Alaska, and the environmental effects of climate change affect human settlements and activity. Alaskan communities are accustomed to adapting to seasonal patterns; however, the unpredictability of the temporal and spatial changes in the environment is creating food security worries. Changes in caribou migration patterns, wildlife ranges, and freeze-thaw events on rivers, among others, are disrupting food-gathering activities and success.[18] People and brown bears are tightly connected in Alaska, so it is inevitable that their interactions will change as climate warming alters the ecosystems they both use.

People preferentially hunt moose and caribou and fish for salmon, foods also preferred by bears. If these resources become less available to people, competition for them may trigger calls for more intensive bear management, such as predator control or liberalized harvest regulations, across the state. It is not clear that climate warming will increase human-bear conflict, but competition over resources could create a conservation concern for bears, so we think this is a key topic for future research.

BEARS ARE HIGHLY ADAPTABLE

Brown bears are very intelligent animals that exhibit flexible behaviors. They live in all of Alaska's major biomes (forest, tundra, and alpine areas) and consume hundreds of different foods. Their widespread distribution and myriad ecological connections make it inevitable that they will be strongly affected by climate change in many and complex ways—some beneficial, some detrimental. Bears have already demonstrated their capacity to adapt to change. This adaptative capacity may allow them to adjust to the environmental effects of climate change through changes in their behavior much faster than would be the case with a less-flexible species. It is impossible to predict whether the net effect of climate change on brown bears will be positive or negative; indeed, some populations may thrive while others decline.

NOTES

1. International Panel on Climate Change, "Climate Change 2021."
2. International Panel on Climate Change, "Climate Change 2021."
3. Box et al., "Key Indicators of Arctic Climate Change."
4. Bhatt et al., "Key Indicators of Arctic Climate Change."
5. Rogers et al., "Thermal Constraints on Energy Balance, Behaviour, and Spatial Distribution of Grizzly Bears."
6. Box et al., "Key Indicators of Arctic Climate Change."
7. González-Bernardo et al., *July Denning in Brown Bears*.
8. Swanson, *Vegetation and Snow Phenology Monitoring in the Arctic Network through 2020*.
9. Swenson et al., "Winter Den Abandonment by Brown Bears Ursus Arctos."
10. Van de Kerk et al., "Environmental Influences on Dall's Sheep Survival."
11. González-Bernardo et al., *July Denning in Brown Bears*.
12. Levi et al., "Community Ecology and Conservation of Bear-Salmon Ecosystems."
13. Prevéy et al., "Warming Shortens Flowering Seasons of Tundra Plant Communities."

14. Hilderbrand et al., "Effect of Seasonal Differences in Dietary Meat Intake on Changes in Body Mass and Composition in Wild and Captive Brown Bears."
15. Tape et al., "Range Expansion of Moose in Arctic Alaska Linked to Warming and Increased Shrub Habitat."
16. Joly, Duffy, and Rupp, "Simulating the Effects of Climate Change on Fire Regimes in Arctic Biomes."
17. Marcot et al., "Projected Changes in Wildlife Habitats in Arctic Natural Areas of Northwest Alaska."
18. Kasischke et al., "Alaska's Changing Fire Regime."

It is impossible to predict whether most climate change effects on bears will be positive or negative—there will likely be a mix of both. Bears have already demonstrated their capacity to adapt to changing conditions. (NPS/Kyle Joly)

6 POPULATION ESTIMATION

HOW MANY BEARS ARE THERE?

Joshua H. Schmidt

A basic question when it comes to brown bears is: how many are there? Although it might sound like a simple matter to just send a biologist out to count them, it is not that simple. Because bears move in time and space, the number of bears the biologist sees is not useful unless we know the geographic area and the time period covered (the survey design), how many were likely missed (detection probability), and our confidence in the resulting number. An appropriate survey design provides the necessary context to ensure that our number is interpretable, valid, and useful. This chapter helps the reader understand the importance of survey design when attempting to answer the question "how many" in a meaningful way.

BASICS OF SURVEY DESIGN

The proper interpretation of survey results relies heavily on the underlying (often unstated) design. The survey design defines the relationship between the population of interest and the population exposed to sampling. If these 2 populations are not completely overlapping, the *abundance estimator* (the specific

◄ Researchers collect all kinds of information on bears, but one of the most common questions seems simple: how many are there? That isn't always easy to answer. (NPS/Nina Chambers)

https://doi.org/10.5876/9781646427116.c006

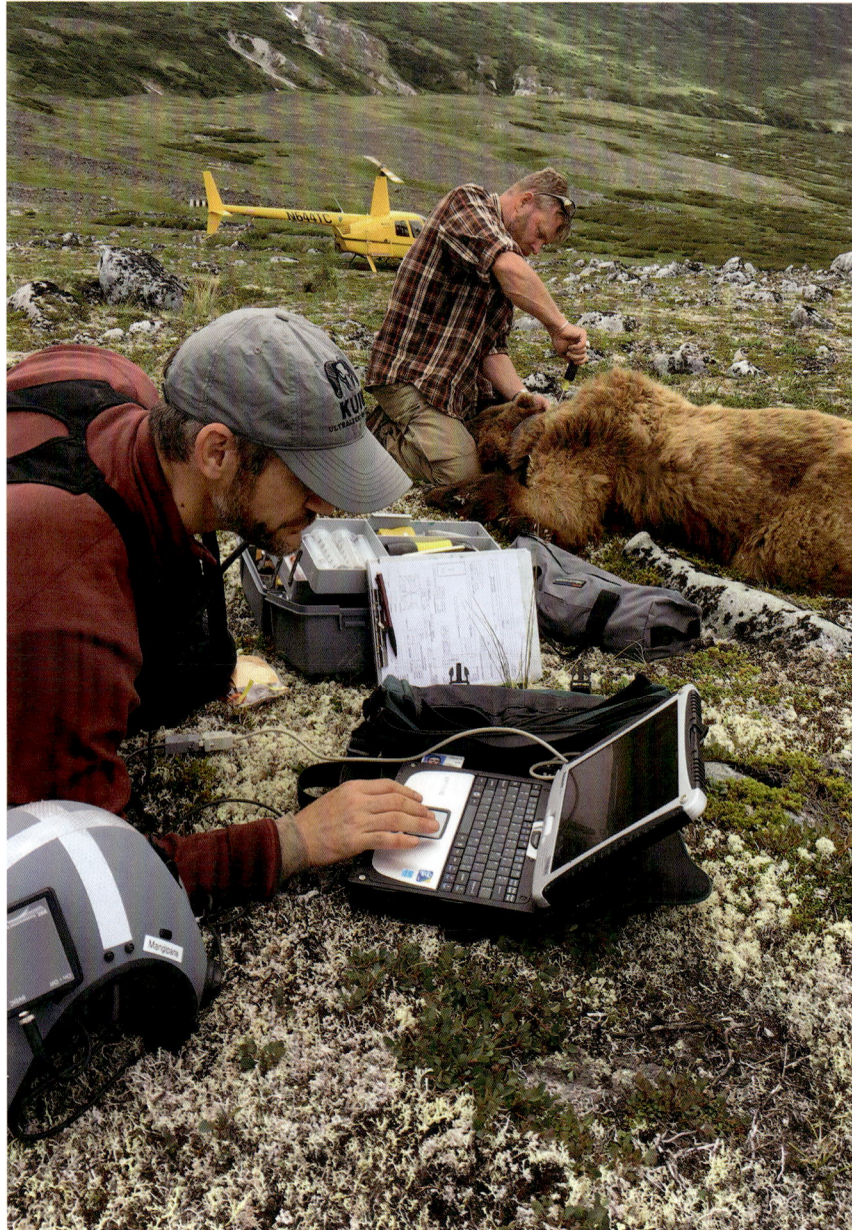

method used to obtain an estimate) may be *biased* (the difference between estimated and true abundance). The first step in designing a survey is to define the population of interest: do we mean "all the bears in the park in the spring," "the number of bears using the Kelly River in July," or something else entirely? We must clearly articulate which group of bears we care about during a particular time frame before we can properly design a survey targeting our population of interest. Let's assume that we are interested in "all the bears in the park" at some particular time when they are not in their dens, such as spring. Park managers often use such estimates to assess the impacts of visitation, harvest, climate change, and other large-scale disturbances on bear populations.

Once we have defined the population of interest, we then need to select a sample that is directly aligned with the population of interest. We cannot cover every square inch of ground searching for bears, so instead we survey a portion of the area (i.e., a sample) that we can use to produce an unbiased estimate of the total number of bears in the entire area. The strongest designs rely on random or systematic distribution of sample units (e.g., survey units, grid cells, transects) throughout the study area to ensure that all individuals in the population

Researchers fit a bear with a collar. We can learn a lot about bears by using collars and tracking their movements. (NPS)

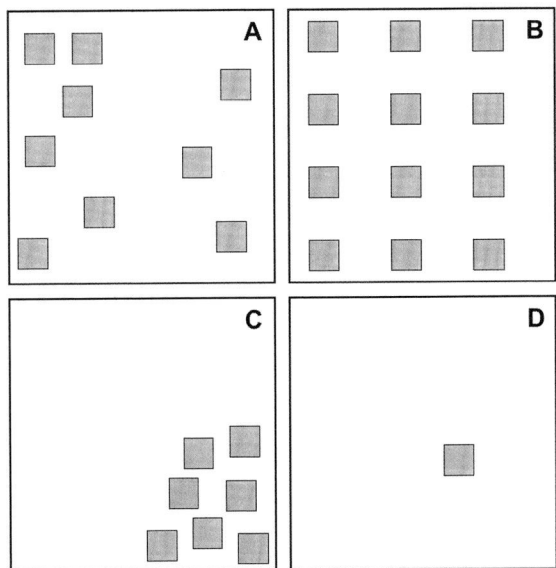

FIGURE 1. Four types of survey design: random (a), systematic (b), non-random (c), and an index site or trend count area (d). Gray cells indicate sampled units, and the surrounding square represents the border of the area of interest. Random and systematic designs provide inference to the entire population within the area of interest (large squares) and require no assumptions about animal distribution. In contrast, the non-random and index site designs only provide inference to the larger area if animal distribution is assumed to be uniform. Given that the distribution of animals is typically unknown and often clumped, random or systematic designs are at much lower risk for bias.

of interest have some chance of being exposed to sampling (figure 1a and d). We can then extrapolate from the sampled area to the entire study area to obtain a valid estimate of population abundance. When a non-random design is used (figure 1c and d), extrapolation over the entire area of interest may not be valid. This is the primary problem with studies that rely on non-random designs (for example, using roads or trails as sample units): the relationship between the sampled population and the target population is undefined. Under a non-random design, one must implicitly assume that the sampled area is representative of the study area: an often untestable assumption.

Once we have determined our basic design, we must then determine our target sample size or how many survey units to cover. Every abundance estimator contains uncertainty, commonly quantified using confidence intervals that tell us how confident we are in an estimate. Sample size has a direct impact on estimator uncertainty and is the main determinant of survey cost. As a simplistic example, say we surveyed just a single sample unit covering 1% of our study area. This would be a very inexpensive survey, but we would not be very confident in an estimate based on this single sample because we do not know how representative it might be of the population of interest. As more and more sample units are added to the survey, we become more confident in the resulting estimate, but cost also increases. The key is to strike some balance between sampling every possible unit (very costly, low uncertainty) and sampling too few units (inexpensive, high uncertainty). Unfortunately, it is impossible to recommend an appropriate sample size for all wildlife abundance surveys because the optimal number of samples depends on the distribution and density of a given population and the methods used to count and analyze the results. However, some helpful guidelines exist based on field sampling approaches.

In addition to a sound design and adequate sample size, we must also confront the fact that during any survey, there will be bears in our sample areas that we do not see. Bears can be difficult to observe because they blend in with the landscape, especially when they are not moving. Without knowing how many bears we do *not* see during a survey, we are unable to put the resulting abundance estimate into perspective. The likelihood of finding and counting a bear that is present and available to be counted is commonly called *detection probability*. Estimating detection probability is a critical component of any wildlife survey, and accounting for it allows us to compare abundance estimates through time and space. For example, say we counted 500 bears during a survey. If we missed 50 bears, our count of 500 is a pretty close approximation of the true population size, but it is not very close if we missed seeing 1,000 bears. The problem is that we have no way of knowing how many bears we *didn't* see if we do not assess detection probability. If we don't know how close an approximation our sample is to the true population, then it is difficult to compare our count with those collected at other points in time

or in other areas. Further complicating the issue is the fact that detection probability often changes between surveys due to factors such as weather, snow cover, observer experience, and other variables. If we count 600 bears in the same area next year, we don't know if there was an actual population change or we just saw more of them (a higher detection probability) during the second survey (better weather conditions may have made bears easier to see). For these reasons, surveys that do not address detection probability directly are generally of limited utility for management, except in cases where changes in population size are very large. Selecting the "best" detection-corrected field approach can help address these problems.

SELECTING A FIELD APPROACH

After thinking through the basics of sampling design, we must determine which field approach is most appropriate for the area of interest. The simplest option is a basic *index survey* whereby bears are counted in some defined area (figure 1d). Index surveys, sometimes called *trend count areas*, are popular because they are typically cheaper and easier to implement than other field approaches. However, given that they usually follow a non-random design consisting of a single or a few sites, they implicitly assume that the counts or trends in the counts over time are representative of the population of interest. As discussed above, this is a tenuous assumption. For example, consider an annual index count of bears fishing for salmon on a particular stream. The assumption is that a consistent proportion of the total population of bears uses the stream so that trends represent changes in the overall population. However, any trend in the counts might be due to a mix of change in the bear population, salmon numbers, detection probability, or a combination of all three. On top of that, there may be other factors (such as the availability of other food resources elsewhere) that influence stream use beyond simple bear population abundance. Unfortunately, there is no way of disentangling these processes without additional information, which may not be available. For this reason, index surveys provide relatively weak inference. A better option is to use a *detection-corrected* survey approach whereby detection probability is incorporated into the abundance estimator.

Two broad categories of detection-corrected survey approaches are commonly used for estimating brown bear abundance at large spatial scales: capture-recapture and distance sampling. Each of these approaches incorporates random sampling, spatial and temporal alignment of sample units with the population of interest, and measure of detection probability—all elements of a strong design.

In brief, *capture-recapture* methods rely on a subset of animals that we "capture" and "mark" using radio collars, DNA samples, or identifying physical characteristics. These marked animals are then "released" either physically or metaphorically, depending on the type of mark used. After waiting some relatively short period of time, we subsequently "capture" another sample of animals and note any animals marked in an earlier period that have been recaptured. Given that we know how many marked animals were released in each period, we can use the proportion subsequently recaptured to estimate detection probability. For example, say we mark and release 10 individuals. We then conduct a survey and see 25 bears, 5 of which are marked. Given that we saw 5 of the 10 individuals we know were marked, we can estimate that we detected about 50% of the individuals present. If our marked sample is representative of the population, then we can estimate that the population size is 50 individuals.

Distance sampling in its simplest form is unique in that it does not require marked individuals; instead, it relies on the idea that animals further away from the observer are more difficult to see. Distance sampling surveys typically consist of a collection of line transects that are searched by an observer who records the distance of each observed animal from the transect line. If we assume that we see all animals that are on or near the transect line (i.e., animals that are very close to us), we can then model the decline in detection probability with increasing distance. The result is a detection function (figure 2) that allows us to understand how detection probability changes with distance. Using the detection function, we can then estimate how many animals we missed at further distances given our reference point of 100% detection on (or near) the transect line. When some bears are missed even at close distances, we can combine components of the mark-recapture and distance sampling techniques as described below.

FIGURE 2. A histogram showing the number of bears detected in each distance class (gray bars) and the fitted detection function (black line) for a hypothetical distance sampling dataset. Given the assumption of perfect detection on the line (i.e., detection probability = 1.0 at distance = 0) and the expectation that there should be equal numbers of bears at all distances, the detection function can be used to estimate the number of bears missed (hashed area) and, hence, total abundance.

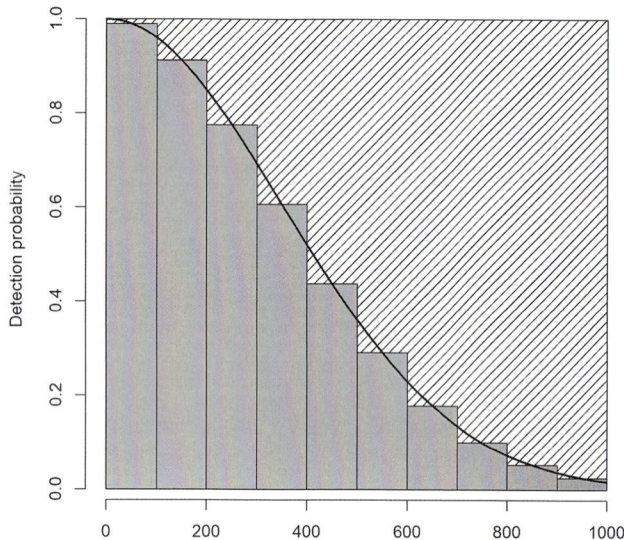

Capture-Mark-Resight

The capture-mark-resight approach has been commonly used for estimating brown bear abundance in Alaska. It combines a radio-collaring project (marking) with an aerial survey (resighting) to estimate bear abundance (figure 3). Researchers used this approach throughout Alaska in the mid-1980s to the mid-1990s and still use it today in some areas.[1] The approach requires at least 2 repeated aerial surveys over the area where marked bears were released. During the survey, pilot and observer teams search for bears from a small airplane, noting any that are marked with a collar. After finishing the survey, the team uses telemetry equipment to check to see if there were any collared bears in the unit that they failed to observe. Based on how often they miss collared bears during the survey, the biologists can then calculate the detection probability for the average bear, marked or unmarked. The total number of observed bears can then be converted to estimated abundance based on the detection probability. The entire area of interest is typically searched during each replicate survey; therefore, surveys do not usually rely on a randomized design.

Collar-mark-resight has provided reliable estimates of bear numbers throughout Alaska over many years. However, the approach has 3 important constraints: (1) the cost of capturing and collaring bears is high, (2) capture and collaring includes risks to both bears and biologists, and (3) the survey area is limited to the area in which bears are captured and released. Therefore, this method precludes valid extrapolation of the results across the larger landscape, but estimators of density and abundance are valid for the defined survey area. For long-term monitoring applications at the landscape scale, the collar-mark-resight approach can be impractical because the costs of replicate surveys and maintaining a collared sample of bears are simply too great.

Distance Sampling

Since the late 1990s, researchers have worked to develop aerial distance sampling approaches for surveying bears that are not reliant on a radio-marked population.[2] These methods have subsequently been applied in many areas in Alaska to estimate bear abundance.[3] Surveys are typically conducted in the spring, after bears leave their dens but prior to leaf-out of deciduous trees and shrubs. During these surveys, a pilot and observer team search a series of randomly distributed line transects from a small airplane, recording any bears seen and their distance from the transect line. Relying on distance sampling theory, the detection function is used to correct for the number of bears missed on the surveyed lines and estimate the number of bears in the searched area. The total number of bears in the area covered by the transects is then extrapolated over the entire area of interest to estimate total abundance. Given the likelihood

FIGURE 3. Wildlife biologist Kyle Joly prepares to dart a brown bear from a helicopter to collar it.

Biologists observed a radio-collared sow with her young cub from the air (bottom). (NPS/Matthew Cameron)

that some bears are missed even at close distances, most applications to date have been based on mark-recapture distance sampling where detections by the pilot and observer are recorded independently and the capture histories for observed bears (seen by pilot only, seen by observer only, seen by both) are then used to correct for the number of bears missed by both the pilot and observer near the line.[4] Recent studies have shown that the number of bears missed near the line has relatively little impact on abundance estimators for existing applications in Alaska, suggesting that standard distance sampling (pilot and observer working together as a team) may be a more practical alternative to mark-recapture distance sampling in these situations.[5]

Aerial distance sampling approaches can be very efficient for estimating bear abundance over large areas. However, around 60–80 detected *bear groups* (the *group* rather than the *individual* is the sample unit for the purposes of estimation) are needed to produce a reliable estimate when using a standard distance sampling approach, and approximately 150 may be required for mark-recapture distance sampling. In areas with high bear densities, these sample sizes may be easily achieved with a modest number of transects, but as density decreases, large numbers of transects must be surveyed to ensure that enough bears are observed to properly estimate the detection function. In many areas, this can mean covering 100% of the study area, a potentially

impractical level of effort. For this reason, aerial distance sampling surveys are typically most effective in areas with relatively high bear densities.

Sight-Resight

In the mid-2010s, researchers developed another approach based on capture-recapture theory that does not require physical capture of individuals.[6] The resulting *sight-resight* survey is essentially a modification of the capture-mark-resight approach that relies on the physical and spatial characteristics of each observed bear group as temporary marks rather than radio collars. In this type of survey, a pilot and observer team search for bears in a series of systematically distributed grid cells (squares placed on a map). When a bear group is detected, the team notes the coloration, group size, and number and size of cubs; takes photographs; and records the group's location. Within a short period of time, a second team searches the same area, recording any bear groups observed using the same procedure (figure 4). At the end of each survey day, the 2 teams compare notes, photographs, and GPS (Global Positioning System) locations to determine which bear groups were detected by both teams. The pairs of sighting histories created by the 2 sets of observers can then be used to estimate detection probability, just as is done in other types of capture-recapture studies. As long as relatively

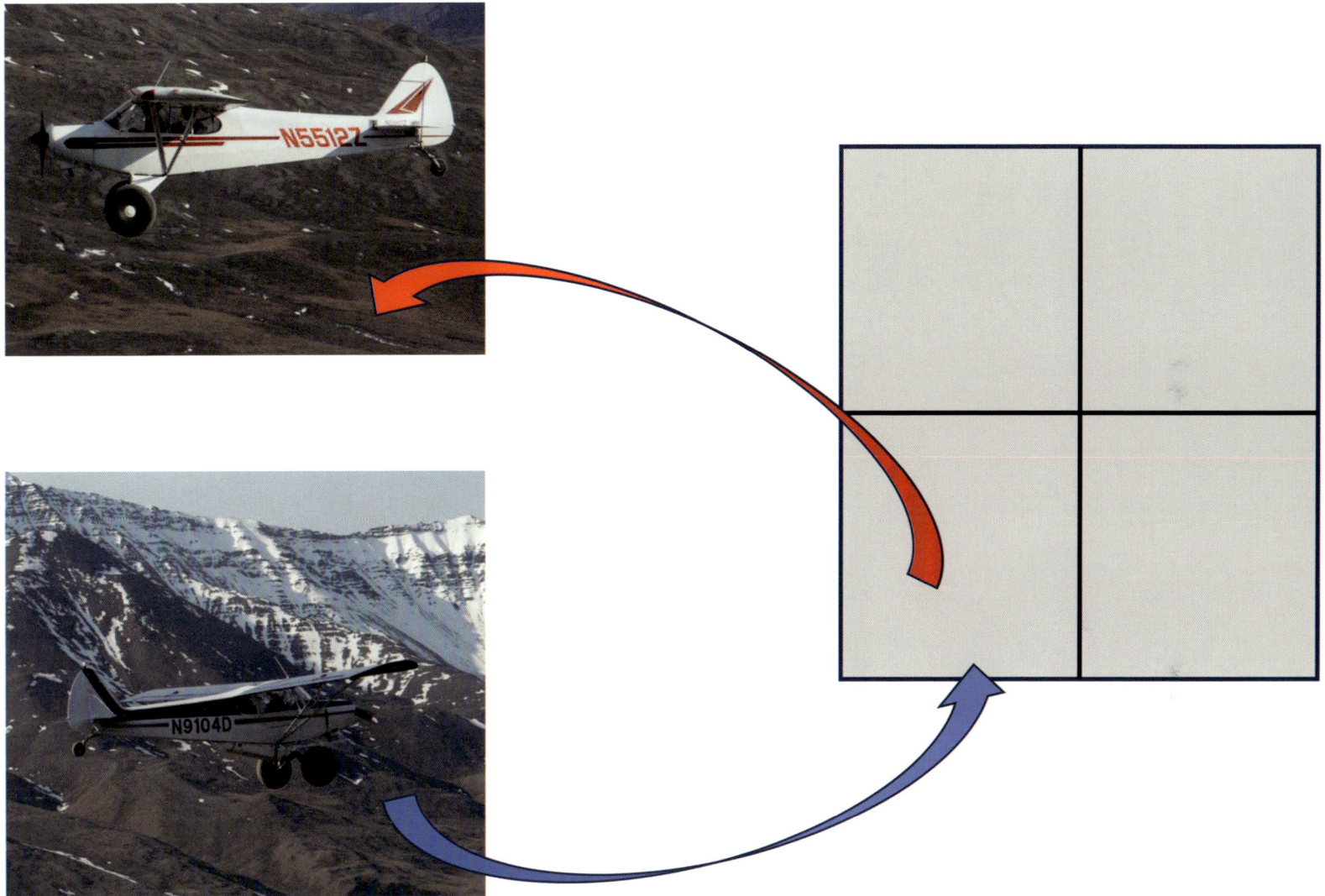

FIGURE 4. Diagrammatic representation of the aerial repeat sampling approach used for sight-resight surveys of brown bears in Alaska. Each subunit (small squares) within a unit (large squares) is searched for bears by both pilot-observer teams in close succession (<4 hours apart). Photographs, spatial location, and other descriptive notes are compared later in the day to identify unique detections, which are then used to estimate detection probability. (NPS)

few bears are present and the 2 survey teams cover the areas in close succession, the temporary marks are sufficient to reliably identify individual bears. This method was developed for use in northwestern Alaska, where habitats are unforested and bear densities are generally low.[7]

As is the case for distance sampling surveys, one advantage of the sight-resight approach is that bears do not need to be physically captured or marked. In a sense, all observed bears can be considered to have been marked over the short interval between repeated surveys. Given that surveys are not restricted to areas containing collared bears, landscape-scale coverage is feasible. Another advantage is that the pilot can optimize the search effort within each grid cell, focusing on areas that are more likely to contain bears or that have more complex habitats. This facilitates a more efficient search pattern than is allowed under a distance sampling survey protocol, resulting in more detections per survey hour. Sample size requirements are roughly similar to those of a standard distance

◄ Photographs depicting the range of group characteristics that can be used to help identify individual bear groups during a sight-resight survey. Note that individual coloration ranges from almost white to dark brown, with a variety of contrasting markings on the back, sides, and legs. Variation in the number, size, and coloration of cubs is also visible. (NPS/Jordan Pruszenski)

sampling survey (60–80 bear group detections), which has been achievable with designs providing 20%–30% coverage of the survey area in the region where this method has been applied. However, the approach is likely less effective in high-density areas, where the identity of individuals may be more easily confused. In addition, the lack of physical marks also requires stronger assumptions about the detection process (e.g., certainty in identification, bears are present and available to be observed by both teams) than are required for the capture-mark-resight approach, where physical marks eliminate ambiguity in individual identification and bears can be located with certainty.

DNA Capture-Recapture/Hair-Snag

DNA capture-recapture, otherwise known as a *hair-snag survey*, is a powerful option used much less frequently in Alaska. This approach relies on the use of DNA identification to mark individual bears and identify any subsequent recaptures. Most commonly, field methods use a series of barbed-wire enclosures throughout the study area with a scent attractant inside. When a bear is attracted to the station and passes through the wire, it (it is hoped) leaves a tuft of hair on the wire that can be processed to "mark" the bear. Some of these marked bears are recaptured at 1 or more stations during subsequent time periods during the survey.

Wildlife biologist Mat Sorum (*left*) and field technician Jared Hughey (*right*) retrieve brown bear hair samples from a hair snare as part of a DNA mark-recapture project investigating use of salmon streams by brown bears in Gates of the Arctic National Park and Preserve. (NPS/Matthew Cameron)

Alternatively, hair samples may be collected from barbed-wire hair snares placed in travel corridors or from specific trees that bears use as scratching posts. Once each individual's hair sample has been identified during each time period, the capture records for each bear are used to estimate detection probability.

These DNA-based hair-snag approaches are most practical when ground access is possible through-out the area of interest. For example, this approach has been used successfully to estimate brown bear abundance in and around Glacier National Park in Montana.[8] In Alaska, hair-snag studies are a less common choice because the lack of road access means that helicopters are typically necessary to place, check, and retrieve stations. Added to the cost of processing DNA samples, total project cost can be prohibitive. However, this method has been used successfully on the Kenai Peninsula, where extensive areas of closed forest make it difficult to see bears from the air, making DNA capture-recapture the only valid option.[9]

PUTTING IT ALL TOGETHER

Given all the possible options and factors to consider, coming up with a rigorous survey design and selecting the "best" survey approach can seem overwhelming. The major deciding factor in selecting among them generally comes down to cost, which is determined by (1) approximate bear density, (2) the spatial scale at which abundance estimates are needed, (3) landscape characteristics (open tundra versus closed forest), and (4) the specific requirements of each method. However, a few broad guidelines may help:

- In general, index surveys should be avoided in favor of a detection-corrected approach.
- In forested habitats, visual survey techniques are simply not practical, although capture-mark-resight may be feasible in areas of discontinuous forest cover. In closed-canopy forest, DNA mark-recapture may be the only practical choice.
- In more open landscapes, capture-mark-resight is a common choice because the ability to find and identify each marked bear during every survey provides solid information about detection probability and relies on fewer assumptions than sight-resight or distance sampling. However, high cost and the limited spatial extent of surveys prohibit inference to large areas. Therefore, this approach may be most practical when used in conjunction with an existing collaring project (see chapter 8) or when inference is desired at more local scales (i.e., ≤2,000 km²).
- In largely unforested coastal areas, where bear density is relatively high due to the proximity of marine resources (Katmai National Park and Preserve, Lake Clark National Park and Preserve, chapters 10 and 11, respectively), distance sampling may be a good choice for providing park-wide estimates. In these areas, higher densities ensure that adequate numbers of bears are detected at a reasonable level of effort. In the large, open habitats of the Arctic parks, where bear densities tend to be low (Gates of the Arctic National Park and Preserve, chapter 9), the sight-resight approach may be a good choice.
- For landscape-scale surveys (>2,000 km²), distance sampling and sight-resight approaches are useful options. Although the type of survey may vary among parks, when designed and conducted properly, density estimates are directly comparable.

FUTURE DIRECTIONS

Survey cost remains one of the main factors limiting the amount of information collected on brown bear populations in Alaska. Although assessing bear populations will likely always be expensive, further development of advanced analytical tools provides one route to cost savings that is often overlooked. When designing wildlife surveys, the standard practice is to treat every survey as if it was the first one ever. If we've already applied a particular method in numerous areas or repeatedly in the same area over time, does it really make sense to assume that we know nothing about detection probability or the population of interest? Probably not. If we instead consider that a survey in similar habitats conducted under similar conditions should have a similar (but not necessarily identical) detection process, we can begin to leverage the vast quantity of information present in existing datasets to increase survey efficiency. Researchers already consider prior information on the detection process when analyzing distance sampling and sight-resight surveys to improve inference.[10] In a similar manner, observed bear distribution can be used to inform abundance estimators. By accounting for spatial patterns of bear locations directly or through the use of spatial covariates (such as habitat type, proximity to a salmon stream), we can better account for the clumped distribution of bears on the landscape and improve abundance estimators.

Similarly, patterns over time can be leveraged by analyzing a series of surveys together. By jointly analyzing data from multiple surveys in the same area, estimates tend to be pulled toward the mean, reducing the effects of sampling error. Consider a situation in which you conduct a survey in a new area where you know nothing about the bear population. With no prior information, the typical analysis assumes that detection probability is somewhere between 0 and 1 (you detect somewhere between 0% and 100% of the bears) and abundance is between 0 and 10,000 bears. The results from this initial survey indicate that detection probability was around 0.5 and estimated abundance was 500 bears. The next time you do a survey you can use the information from your first survey to effectively say, "I'd be surprised if detection probability was near 0 or 1; in fact, I think it's closer to 0.5." Similarly, "given that only a year has passed since the last survey, I'd expect abundance to still be in the neighborhood of 500, and I'd be surprised if there were fewer than 300 or more than 700 bears." By formally incorporating information from prior surveys in this way, you need to collect less new data (less survey effort is needed) to estimate abundance for the latest survey. Obviously, there are limits to the benefits you might expect, but considering analyses in this way is essentially a way to reduce survey effort for free.

NOTES

1. Miller et al., "Brown and Black Bear Density Estimation in Alaska."
2. Becker and Quang, "A Gamma-Shaped Detection Function for Line-Transect Surveys"; Becker and Christ, "A Unimodal Model for Double Observer Distance Sampling Surveys."
3. Becker and Crowley, "Estimating Brown Bear Abundance and Harvest Rate on the Southern Alaska Peninsula."
4. Becker and Quang, "A Gamma-Shaped Detection Function for Line-Transect Surveys"; Becker and Christ, "A Unimodal Model for Double Observer Distance Sampling Surveys."
5. Schmidt et al., "Improving Inference for Aerial Surveys of Bears."
6. Schmidt et al., "Using Non-Invasive Mark-Resight and Sign Occupancy Surveys to Monitor Low-Density Brown Bear Populations across Large Landscapes."
7. Schmidt et al., "Using Non-Invasive Mark-Resight and Sign Occupancy Surveys to Monitor Low-Density Brown Bear Populations across Large Landscapes"; Schmidt et al., "Brown Bear Density and Estimated Harvest Rates in Northwestern Alaska."
8. Kendall et al., "Using Bear Rub Data and Spatial Capture-Recapture Models to Estimate Trends in a Brown Bear Population"; Kendall et al., "Grizzly Bear Density in Glacier National Park, Montana."
9. Morton et al., "Estimation of the Brown Bear Population on the Kenai Peninsula, Alaska."
10. Schmidt et al., "Improving Inference for Aerial Surveys of Bears"; Schmidt et al., "Brown Bear Density and Estimated Harvest Rates in Northwestern Alaska."

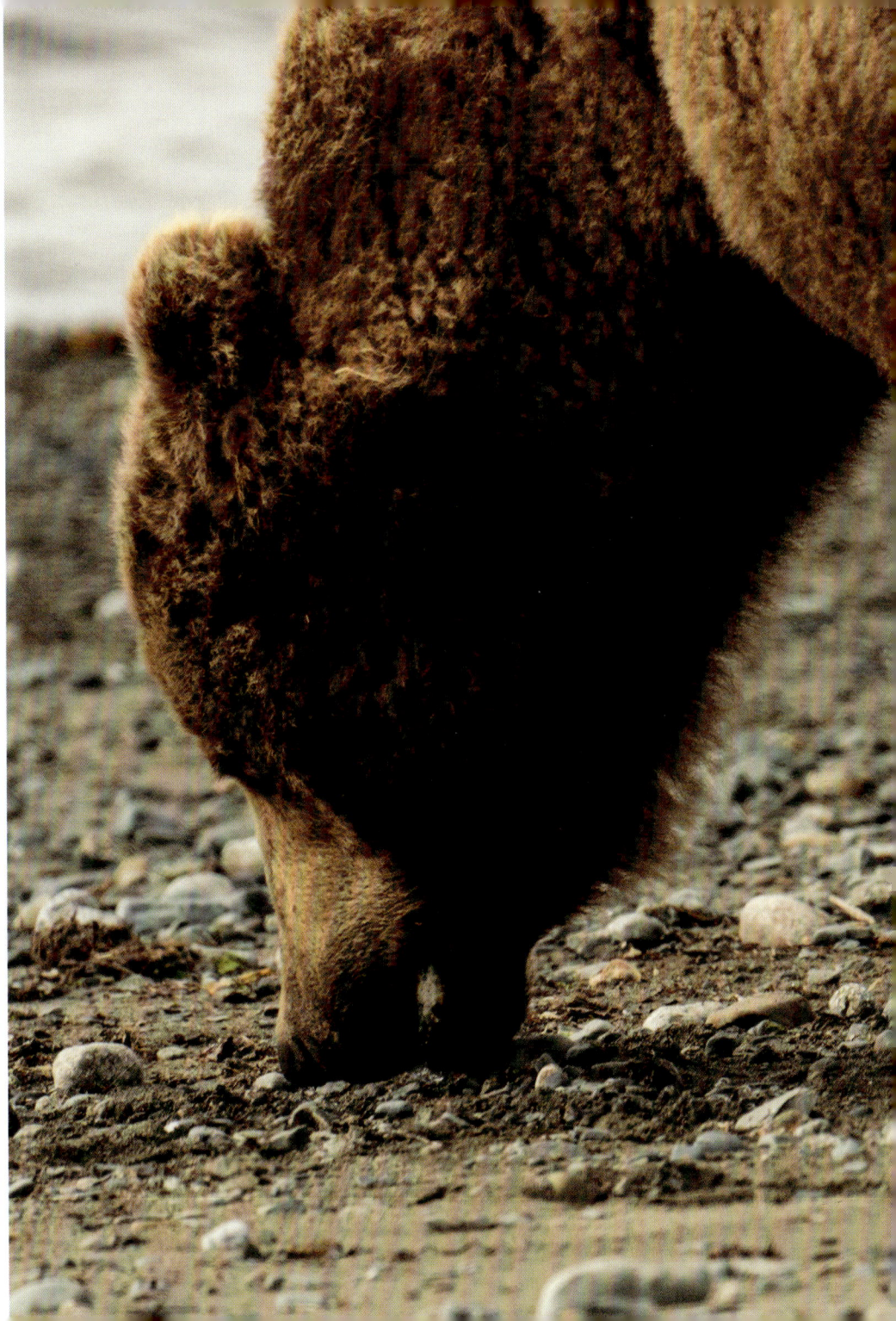

A bear nuzzles the sand for potential food. (Stuart Leidner)

Brown bears expanded into Glacier Bay National Park and Preserve as newly deglaciated landscapes provided habitat. (NPS/Tania Lewis)

7
GLACIER BAY

BEARS IN THE BAY THAT GLACIERS BUILT

Tania Lewis and Mary Beth Moss

Glacier Bay National Park and Preserve encompasses the traditional homelands of the Huna and Yakutat Tlingit, who have sustained themselves for generations on the rich abundance of the area's lands and waters. Glacier Bay's original people not only coexist with but draw sustenance, knowledge, and power from their interactions with both living and non-living beings inhabiting the landscape, including the brown bear, or *Xóots*.[1]

Although subsistence use of brown bears has waned, Tlingit hunt bears for meat, fat, and hides for blankets. In times past, they used the bones and claws for tools and regalia.[2] Hunters and their families traditionally followed carefully prescribed protocols both before and during a hunt to ensure success, and they approached a hunt with humility.[3]

More important, brown bears are kin to their Tlingit brethren. In the Raven story cycle, animals were once humans who were frightened when Raven let daylight loose from its box. Bears and humans, then, are relatives, capable of crossing into each other's worlds. In the ancestral past, a young Tlingit girl

https://doi.org/10.5876/9781646427116.c007

married a brown bear,[4] a surviving clansman invited bears to feast at a memorial party for his departed relatives,[5] and brown bears taught starving Huna clans to dig for cockles during the Year of Two Winters.[6] When traveling in Tlingit country, the Tlingit address the bears as brothers, asking for safe passage through their territory. Women hold special powers over brown bears and can calm them by speaking gentle Tlingit words. The Chookaneidí Clan of Glacier Bay shares a special relationship with *Xóots*, claiming the brown bear as their crest and naming their clan houses after their kin *Xóots Hít* (Brown Bear House) and *Xóots Saagí Hít* (Brown Bear's Nest House), as do the Kaagwaantaan Clan and several other Tlingit clans.

The first Europeans to chart the mouth of Glacier Bay sailed with Captain George Vancouver on the HMS *Discovery* in 1794. At that time, Glacier Bay was merely a slight indentation in a glacier face that was 4,000 feet (1,200 m) thick and extended 20 miles (32 km) across and 100 (161 km) miles back. Almost 100 years later, naturalist John Muir explored the region and found that the glacier had retreated 48 miles (77 km), creating the newly emerging fjord. Brown bears, along with most other mammals, were rare in this newly deglaciated landscape. Muir spread the word about the scientific and spiritual significance of Glacier Bay through newspaper and magazine articles as well as public lectures, sparking the beginning of boat-based tourism. Steamship cruises to the glacier faces became popular throughout the 1880s and 1890s. Scientists, also inspired by Muir's accounts, caught rides on these steamships and began conducting some of the foundational studies of tidewater glaciers and glacial geology, which in turn inspired more research. In 1899, the Harriman Expedition explored Glacier Bay with a crew of many researchers with diverse scientific backgrounds, resulting in the collection of plant and animal specimens, including new species. That fall, a series of large earthquakes caused such a massive calving of icebergs that steamships were unable to get close to the glacier faces the following summer. Steamships began to go to other glaciers in southeast Alaska and never resumed bringing tourists to Glacier Bay, thus ending the first phase of tourism and scientific exploration.

Scientists soon began chartering their own vessels to Glacier Bay, and research on glaciers and postglacial plant succession soon resumed. One scientist, William S. Cooper, presented his findings to the Ecological Society of America in 1922, inspiring fellow ecologists to form a committee to work toward preserving Glacier Bay for scientific exploration—which led to the establishment of Glacier Bay National Monument in 1925. The monument was expanded significantly in 1939 to include brown bear habitat along the Gulf of Alaska. For many years, the monument supported science, mining, fox farming, agriculture, and hunting. Sightings of brown bears

Glacier Bay National Park and Preserve encompasses over 3 million acres in the northern portion of southeast Alaska.
It is characterized by deep fjords, mountains, islands, and iconic tidewater glaciers.

were rare in Glacier Bay in the early 1900s, but near the mouth of Glacier Bay brown bears, possibly some of the early post–ice age colonizers, were reportedly shot by homesteaders to protect cattle. In the late 1960s, boat-based tourism resumed, with visitors enjoying tidewater glaciers on vessels ranging from kayaks to ships; by then, brown bears were more commonly seen on the shoreline in Glacier Bay. In 1980, the monument was expanded again under the Alaska National Interest Lands Conservation Act (ANILCA) and designated as Glacier Bay National Park and Preserve, world renowned for vessel-based viewing of glaciers and marine and terrestrial wildlife, including brown bears.

ENVIRONMENT

Glacier Bay National Park and Preserve encompasses 3.34 million acres in northern southeast Alaska. The current climate is characterized by cool summers and wet winters.

Temperatures average around 50°F–60°F (10°C–16°C) in the summer and 25°F–40°F (−4°C–5°C) in the winter. Rainfall averages around 60–70 inches (152–178 cm) per year, and snow depths and duration vary widely by elevation and proximity to ice fields. The topography consists of mountains up to 15,325 feet tall (4,671 m; Mount Fairweather), ice fields, glaciers extending to tidewater, and glacially carved mountains and valleys.

The deep fjords, islands, and mountains that characterize southeast Alaska were shaped by a mixture of geological and tectonic processes, with dramatic glacial periods of advance and retreat. Around 280 years ago, Glacier Bay was largely covered in ice during the Little Ice Age, while portions of the park's outer coast remained ice-free, providing refugia to land plants and animals. Since then, the glaciers have receded at an unprecedented rate, exposing 50–60-mile-long (80–97 km) fjords and freshly sculpted land. Plants and animals colonized Glacier Bay as the glaciers receded. As mammals migrated into the area, the glaciated and mountainous landscape and wide fjords geographically influenced the colonization by acting as barriers and directing the flow of migration.

Glacier Bay currently encompasses a wide spectrum of habitat, ranging from barren glacial outwashes near the glacier faces to rich old-growth forest in refugia areas not glaciated during the Little Ice Age. This wide range of terrestrial and shoreline habitats contributes to diverse and abundant food resources for brown bears in the park. Early pioneer-stage plants offer few food resources to bears, but young open-scrub habitats often contain a mosaic of bear foods including soapberry, strawberry, locoweed, and bear root. Closed-scrub habitats dominated by willow often contain extensive soapberry, while alder-dominated closed scrub contains little high-quality bear forage with the

exception of groundcone, an alder root parasite. Young forests in the southern portion of the Glacier Bay are dominated by dense Sitka spruce with little understory. Several berry-producing species including blueberry, salmonberry, and red elderberry grow in young forest openings and fringe. Areas surrounding Glacier Bay that were not glaciated during the Little Ice Age consist of old-growth hemlock and spruce forests with interspersed peat muskegs. Old-growth communities often contain dense understories of berry-producing plants including blueberry, salmonberry, and devil's club berries. Skunk cabbage is common in wet areas, and low-bush blueberries and many herbaceous plant

Deep fjords, islands, and mountains were shaped by glacial processes, creating an abundant coastal habitat for bears. (NPS/Tania Lewis)

Diverse habitats provide plants and berries for bears as well as moose, mountain goats, small mammals, bird eggs, and marine life. (NPS/Tania Lewis)

foods are present in alpine areas. Riparian zones host important spring sedge habitat and seasonal anadromous fish runs that increase in diversity and abundance with the number of years since glaciation.[7] A large portion of the coastline throughout the area is bordered by recently uplifted grass and herbaceous beach meadows. Bear food resources in these meadows include strawberry, nagoonberry, dandelion, horsetail, angelica, cow parsnip, beach lovage, pacific hemlock-parsley, and grasses. In addition to the diversity of bear plant foods available, inter-tidal food sources including barnacles, mussels, and rock gunnels appear to make major contributions to bears' diet, especially in more recently deglaciated areas where plant food resources are more minimal. Other potential prey species throughout the study area include moose, mountain goats, voles, ground-nesting bird eggs, marine mammals, and invertebrates such as amphipods, ants, bees, and wasps.

The majority of park visitation in Glacier Bay proper occurs by motorized vessel. Of the roughly 550,000 visitors per year, more than 95% come by cruise ship and never set foot on the ground within the park. Approximately 600–1,600 visitors camp along the shoreline of Glacier Bay every year. Day use of the park's shoreline from tour vessel shore excursions increased from 1,246 people in 2016 to 3,634 people in 2021.[8]

MANAGEMENT CHALLENGES

Human-Bear Conflicts

Historical accounts of human-bear conflict in Glacier Bay date back to 1912 when Allen Hasselberg, later known as the "Bear Man of Admiralty Island," was mauled by a brown bear as he was hunting up the Bartlett River.[9] He later collected the skull of the bear that mauled him for C. Hart Merriam, lead zoologist of the Harriman Expedition, who designated the specimen as a unique species, *Ursus orgilos*, but that was later determined to be *Ursus arctos*. Miners in Glacier Bay and settlers raising cattle in Gustavus in the 1920s to 1950s were challenged by "marauding" brown bears that, according to homesteaders' recollections, were shot on sight. Reported human-bear conflicts were minimal (1–2 per year) and minor from 1960 through 1975. In 1976, a lone kayaker camping in the east arm of Glacier Bay was killed and consumed by a brown bear. In 1978, our Glacier Bay National Monument staff wrote the first bear management plan and made attempts to bear-proof garbage cans and the dump in Bartlett Cove to deal with an estimated 25 black bears in the frontcountry that were partially or entirely dependent on human food sources. The plan defined methods to reduce human-bear conflicts as well as to protect and maintain natural habitat for black and brown bears. After the first bear management plan was written, the number of human-bear conflicts continued to increase, and another lone kayaker was killed by a black bear in Sandy Cove in 1980. These events led to repeated yearly seasonal camping closures of 2 large sections of coastline: Sandy and Spokane Coves and West Tarr/North Johns Hopkins Inlet (referred to from here on as the Sandy Cove and Tarr Inlet closure areas). In addition to the camping closures, these areas were subject to periodic monitoring of bear distribution, habitat, and abundance throughout the 1980s.[10]

In 1988, Glacier Bay, now a national park, completed its second bear management plan, which further detailed methods used to reduce human-bear conflict while preserving bears and their habitat and allowing for visitor education and enjoyment. Although the bear management plans of 1978 and 1988 outlined specific methods of eliminating all human food sources for bears as an important step in reducing human-bear conflict, bears continued to obtain human food and trash in the backcountry throughout the 1980s and into the early 1990s. In 1991, we began to mandate the use of bear-proof food storage techniques such as bear-resistant food containers (BRFC) in the backcountry, and incidents in which bears obtained food dropped over the next few years. In the frontcountry, bears continued to get food regularly, despite the bear-proof garbage cans, until the summer of 1992 when a black bear with 3 new cubs repeatedly obtained food from peo-

ple in Bartlett Cove and was subsequently relocated to Geikie Inlet. After that season, food storage in the developed area improved significantly, and overall numbers of conflicts decreased throughout the park for several years.

Human-bear conflict numbers began to rise again with increases in backcountry users in the late 1990s and peaked at 22 in 2000. Most of the conflicts during this time involved brown bears in the backcountry. Backcountry visitation peaked at more than 1,500 campers per season in 1997 and has decreased to less than 1,100 per year since 2005. Increases in backcountry use and bear-human conflicts in the late 1990s led to the initiation of several bear research projects during the years 2001–2005 designed to minimize those conflicts, evaluate the ongoing Sandy Cove and Tarr Inlet closure areas, and inform a comprehensive bear management plan. Lessons we learned in the first few years of this phase of bear research led to an overhaul of the park's bear-safety message beginning in 2003. The new safety message taught campers how to interpret basic elements of bear behavior and how to react accordingly during bear encounters, and it encouraged people to carry bear pepper spray, maintain control of their gear, and stand their ground against approaching bears in most situations.[11] Since the new safety message was instituted, the park has had approximately 6–7 human-bear conflicts per year.[12]

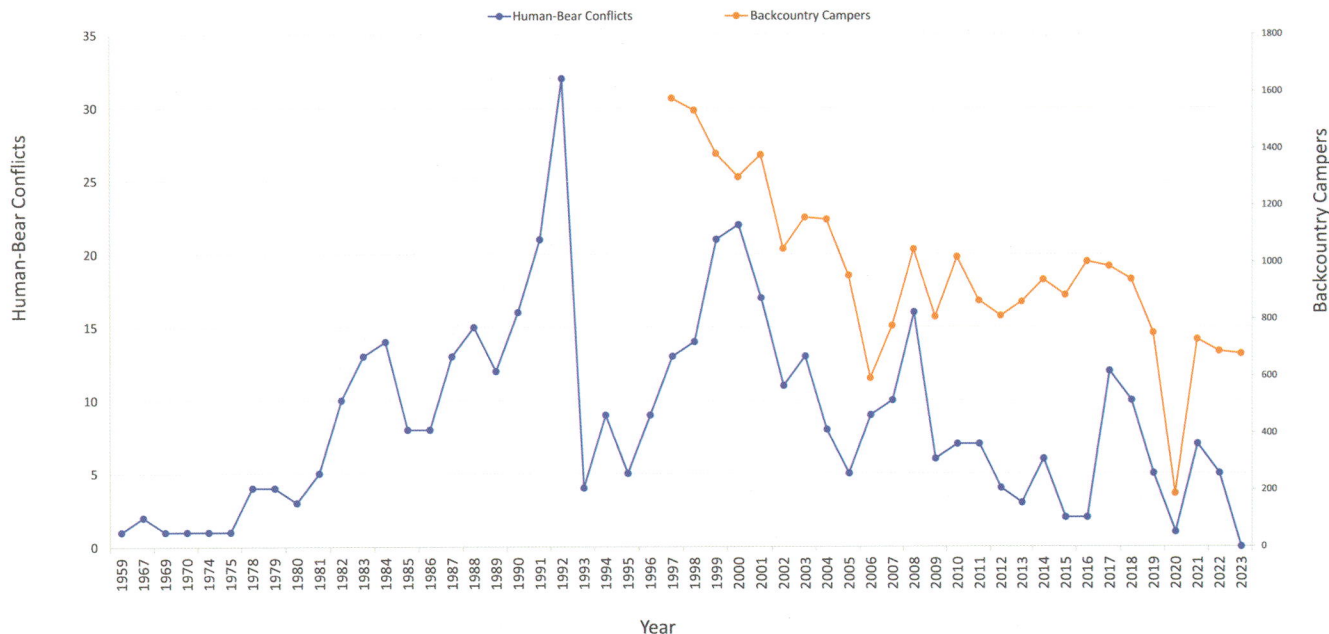

Number of human-bear conflicts since 1959 and number of backcountry users since 1997 in Glacier Bay National Park and Preserve, by year.

Bear Viewing

Some of the biggest draws for visiting Glacier Bay are glaciers and wildlife, including bears. Almost everyone who views bears in Glacier Bay does so from vessels. To study the effects of vessel-based bear viewing on the behavior of brown bears in Glacier Bay during the years 2008–2010, we experimentally approached 24 brown bears from motorized vessels and recorded the distance from boat to bear using rangefinder binoculars. During these 1- to 10-minute approaches, the bear's behavior was documented every 15 to 30 seconds. We categorized bear behaviors as either energetic gain (beneficial, such as feeding and resting) or stress (detrimental, such as vigilance and mouthing) behaviors. Our results indicate that energetic gain behaviors did not change significantly with boat proximity, but the frequency of stress behaviors increased significantly when a vessel approached a bear within 100 yards (100 m).[13] In addition, the majority of bears that were approached within 100 yards (100 m) fled short distances, though several bears were completely displaced from the beach. Our findings led to outreach material for boaters in Glacier Bay, with recommendations to stay at least 100 yards (100 m) from bears on the beach and to watch for disturbance behaviors that might indicate that one's vessel was too close.[14]

Almost everyone who engages in bear viewing in Glacier Bay National Park and Preserve does so by boat. Of the approximately 550,000 visitors to the park each year, 95% come by cruise ship and never set foot on land. But approximately 600–1,600 visitors camp on the beach. (NPS/Tania Lewis)

Harvest and Defense of Life and Property

Sport hunting of both brown and black bears occurs in Glacier Bay National Preserve, commonly called Dry Bay, at the mouth of the Alsek River. Harvest data show that between 3 and 14 brown bears are harvested from the preserve each year, with an average of 5.8 since 1960.[15] Although Dry Bay is remote, there are several airplane landing strips that hunters use as well as a public-use cabin in which hunters can stay. There is also a guide in residence in Dry Bay who specializes in guided bear hunts, both in the preserve and on adjacent United States Forest

Service lands. There have been 3 reported defense of life and property (DLP) kills in the preserve, all associated with bear hunters, as well as numerous anecdotal reports over the years of nuisance bears killed in Dry Bay illegally. Park staff use education and enforcement in Dry Bay to attempt to minimize DLPs and illegal nuisance bear kills.

A recent unpublished study by scientist Anthony Crupi found that 2 male brown bears captured and collared by the Alaska Department of Fish and Game (ADF&G) at the Yakutat landfill migrated to Dry Bay and back to Yakutat, where they were subsequently killed. One of these bears denned for a winter south of Dry Bay within the national park. These data provide the first evidence that actions taken in Yakutat affect park resources. Harvest and DLP kills in the town of Gustavus, immediately adjacent to the park, also have the potential to affect bear populations, but the overwhelming number of DLP and sport hunting kills involve black, not brown, bears.

RESEARCH SUMMARY

Brown Bear Habitat and Activity

Increasing human-bear conflicts, including 2 fatalities and multiple incidents involving bears obtaining campers' food, led to repeated seasonal camp closures in the 1980s and bear monitoring in the late 1980s to attempt to gain information on bear habitat in areas of concern. Our first comprehensive study of bear habitat and human-bear encounter potential was conducted in 2001–2002. We assessed 161 camping areas along the shoreline of Glacier Bay for bear habitat potential, bear encounter potential, and bear displacement potential. Subjective encounter risk ratings were assigned to each site in the field, and a statistical analysis was conducted to test which site variables are most indicative of bear activity levels. Seventy-three percent of those campsites were subjectively rated as moderate bear encounter risk, 20% were rated as low encounter risk, and 7% were rated as high encounter risk. Statistical analysis involved deriving human use levels for specific areas from the park's camper database and identifying relative bear usage and bear-human conflict histories for specific areas by constructing a database of park records. This study found that the probability of human-bear conflict was best predicted by the following variables, in decreasing importance: distance to salmon streams, amount of bear sign, and bears' ability to bypass the site.[16]

In 2004, we collaborated with the United States Geological Survey on a study to quantitatively evaluate the 2 large (17 miles and 18 miles [28 km and 29 km], respectively) sections of shoreline in Sandy Cove and Tarr Inlet that had been closed to the public for camping for more than 20 years due to human-bear conflicts. Park managers needed

Non-Invasive Wildlife Research

Non-invasive wildlife research is a term used for wildlife studies that do not involve capturing animals. Recent advances in technology allow biologists to learn more and more about animals by simply collecting their hair, scat, or photographs. DNA extracted from hair can be used to identify species, sex, individual identity, and genetic population structure of animals in a certain area. Hair can also be used to evaluate components of diet for the period the hair grew, using stable isotope analysis. Bears are especially easy to collect hair from because of their tendency to rub against trees and investigate interesting smells. Scientists use these behaviors to their advantage by securing barbed wire on bear rub trees or creating a corral of barbed wire around a strong-smelling liquid. As bears rub or investigate smells, they often leave a discrete hair sample that can be analyzed. Scat can also be analyzed visually for contents as well as genetically (if fresh) to identify individuals and identify the food items in the scat.

Remote cameras (often called game cameras) can capture time-lapse and motion-sensor–activated still and video images that can be used for conducting behavioral studies, assessing occupancy and activity levels, and identifying individuals with distinct markings. Identifying individuals with repeat sampling using any of these methods can be used to generate population estimates using capture-mark-recapture techniques. With non-invasive techniques, the animal is "captured" by being identified genetically or on camera, without ever being touched by the researcher. These methods are especially appropriate in many wilderness areas of Alaska national parks, where capturing and collaring animals is considered a degradation of wilderness character.

information on bear use in these areas to determine whether the long-term camping closures remained warranted. They also needed information on effective methods for evaluating bear activity and habitat quality for future conflict sites. We tested several non-invasive methods to assess bear activity and habitat quality within these closure areas as well as within other areas of historically high numbers of human-bear conflicts. In June, July, and August of both 2004 and 2005, we conducted biweekly surveys of 8 study areas (6 predominantly brown bear areas and 2 black bear areas) to quantify bear activity using remote time-lapse videography, bear sign abundance surveys, and genetic testing of hair samples.[17] We also evaluated bear habitat quality by mapping and classifying habitats and determining the abundance and diversity of bear foods using vegetation plots on random transect lines within each habitat type. One camping closure area ranked consistently low in bear activity and habitat quality, while the other ranked consistently high. These results led park managers to open the closure area with low bear activity/habitat quality (Tarr Inlet) starting in the 2007 visitor season, while the closure containing high bear activity and habitat quality (Sandy Cove) remained closed to camping until 2015, when a repeat study found that bear use of the shoreline had decreased. This work suggests that habitat evaluation, bear sign mapping, and

Non-invasive study methods do not involve handling animals but rather use hair, scat, and photographs of animals. In this series of photos, we used a camera to identify rub trees, then installed snares to collect hair samples. From the hair, we can use genetics to describe the population. (NPS/Tania Lewis)

periodic scat counts provide a useful index of bear activity for future sites of interest.[18]

Brown Bear Distribution and Genetics

Brown bears range across the southeast Alaska mainland and many of the northern islands, including Admiralty, Baranof, and Chichagof (ABC Islands). Species distribution in the park has varied widely over the past 100 years. In the 1920s and 1930s, brown bears were regularly reported in Gustavus and Bartlett Cove yet were essentially absent in those areas from the 1960s through the early 2000s. Brown bears have since returned to this area and are reported often. A similar increase in brown bears is noticeable since 2010 in western Dundas Bay, at Point Carolus, and in Bartlett Cove and the lower Beardslee Islands.[19]

During the summers of 2009 and 2010, we used direct observations, track identification, and genetic analysis of non-invasive bear hair collections to examine the distribution of black and brown bears across the shoreline of the park relative to the stage of succession of the landscape using an occupancy modeling framework that accounted for differences in detectability.[20] Through this work, we found that black bears are most closely associated with closed forest cover in the southern two-thirds of the bay and are essentially absent from recently deglaciated (<150 years since deglaciation) habitats. Over time, the distribution of black bears will likely move northward as forest develops in newly deglaciated areas. Brown bears were documented at all study sites, with the highest levels of activity occurring in recently deglaciated areas of open scrub and old-growth forest outside Glacier Bay proper and lowest levels of activity in the young forests of southern Glacier Bay.[21] The current distribution of black and brown bears in the park will undoubtedly change as plant and stream communities continue to mature.

In a parallel study from 2011 and 2012, we assessed the bear population in a 124-mile (200-km) study area of the Gustavus Forelands, including parklands and the city of Gustavus, using genetic analysis of bear hair from bear rub trees and scented hair traps. Over the course of the study, we collected 196 hair samples from 25 rub trees and 8 baited hair traps, identifying 33 individual black bears and 14 individual brown bears.[22] The number of individual brown bears identified in the Gustavus Forelands was surprising because they had rarely been seen and have only recently begun using the area after a 50-year absence, providing further evidence of brown bears' recent range expansion.[23] Information from both studies led to increased community outreach regarding brown bear behavior and safety at park headquarters and in the town of Gustavus.

Changes in bear distribution over time are likely influenced by a range of factors, including receding glaciers providing access to new territory, subsequent plant and stream succession, immigration of individuals through travel corridors, and colonization of new areas. Competition between brown and black bears has been hypothesized to play a major role in colonization success when there is a large dietary overlap between the species.[24] Evidence indicates that brown bears have an advantage over black bears when high-quality foods are concentrated at predictable times (e.g., salmon runs), allowing brown bears to dominate foraging through interference, resource defense, or competition. When food resources are more dispersed and less predictable, however, black bears, with smaller body size (fewer dietary requirements) and higher densities, are able to dominate foraging opportunities through exploitation competition. This could explain why brown bears are less prevalent in the lower forested areas of Glacier Bay, where berries and other plant resources are dispersed, and higher densities of black bears are able to exploit the dispersed food resources. Exceptions may occur at salmon streams, where brown bears are able to dominate a high-quality resource.

In 2011–2012, brown bear hair collected from non-invasive research projects in the park, as well as harvest samples obtained from outside the park by the ADF&G, were further analyzed to examine the genetic population structure of brown bears across northern southeast Alaska. Micro-satellite genetic analysis identified 105 individual brown bears. Genetic and landscape analyses were used to examine how the landscape and population structure of brown bears are intertwined in Glacier Bay National Park and Preserve and to help determine likely sources for brown bear recolonization in the recently deglaciated region.

Through this analysis, we identified 3 genetically distinct groups of brown bears and found that both the rugged Fairweather Range and the wide fjord of Glacier Bay are barriers to dispersal.[25] Two genetic groups range far beyond the park's boundary to both the west and the east, and the third group has been isolated long enough to undergo genetic drift and develop a genetic signature unique to northern Glacier Bay. This endemic subpopulation likely stems from an original group of colonizers from the east, while the other 2 groups are more recent immigrants. One recently immigrated group likely moved into the bay from the northwest, while the other arrived from the northeast. Both represent a second wave of colonization along the shoreline of Glacier Bay. These more recent immigrants will likely mix with the original colonizers after years of separation, and the unique genetic signal of the original colonizers may vanish.

Brown Bear–Wolf Scavenger Study

A humpback whale carcass was found washed ashore on the west arm of Glacier Bay in the spring of 2010, so we and our collaborators installed and maintained remote cameras at the site from May 19 to September 17 to document the use of the carcass by scavengers. Both brown bears and wolves were present at the whale carcass repeatedly throughout the entire summer, appearing to tolerate each other at this abundant food source. As many as 6 brown bears at a time fed on the carcass, with the highest activity during the mornings and evenings. Up to 7 wolves were observed, with activity highest in the early morning, and 4 wolf pups joined their parents at the feast in early August. We analyzed photos of what ended up being one of the longest scavenger events of a single carcass ever documented and a unique opportunity to observe wolves and bears scavenging simultaneously.[26]

In September 2010, most of the remaining carcass washed from the Scidmore Cut and drifted south with a brown bear on top of it. The next spring, the carcass was discovered again north of the mouth of Geikie Inlet, and a remote camera was established at the site. The pack of wolves from 2010 continued to feed on the carcass throughout the summer of 2011 in the new location, as evidenced by repeated photos of wolves from the motion-sensor camera, a heavily worn canine trail

Brown bears and wolves scavenge whale carcasses that occasionally wash up on the beach. (NPS/Tania Lewis)

along the beach fringe, and wolf hair and teeth marks on the whale flesh and bones. As the flesh of the whale was reduced, there was increasing evidence of wolves eating the bones, particularly the round ends of ball joints. Brown bears also visited the carcass over the summer, as did a single black bear, but with less frequency than wolves. Once the soft tissue of the whale was completely consumed, scavengers continued to consume bone and marrow out of ribs and large vertebrae. A large brown bear was chewing on bones on November 6, and the pack of wolves with glossy thick winter coats scavenged on the bones as late as November 27. In the spring of 2012, scavengers were no longer reported frequenting the carcass site, and only a few bones remained on the beach.[27]

Effects of Shoreline Tourism on Wildlife Activity

Shoreline tourism has been growing in Glacier Bay National Park and Preserve, largely due to increasing day use by passengers from tour vessels. The shorelines of Glacier Bay provide denning locations, foraging habitats, and nesting habitat for wildlife and see the highest concentration of human use, therefore increasing the frequency of human-wildlife interactions. These interactions can cause wildlife to alter their activity patterns or be displaced; as a result, the interactions impact their survival and reproduction. Therefore, there is a need to quantitatively understand how increased human use of the shoreline is affecting the capacity of wildlife and humans to coexist. Wildlife have varying responses to disturbances depending on the duration, severity, and type of disturbance event. Some disturbances modify wildlife habitat and can impact community assembly and patterns of diversity, while others can modify wildlife behavior. During the years 2017 and 2018, we collaborated

A study of the effects of shoreline tourism on wildlife activity used remote wildlife cameras to quantify the presence and activities of brown bears, black bears, moose, and wolves. The results from cameras from different locations were compared to determine activity level and human presence. (Reconyx remote camera)

with the University of Washington to investigate the behavioral responses of terrestrial mammals, including brown bears, to increased human activity on the park's shoreline areas and the community-level responses of terrestrial mammals to the dramatic glacial recession that formed Glacier Bay. The goal of this study was to provide information to inform park management of the impacts to wildlife occurring across differing levels of human use in an effort to establish biologically relevant thresholds of human use to ensure that significant resource degradation does not take place. In collaboration with the University of Washington, we used remote cameras to quantify the presence and activity levels of brown bears, black bears, moose, and wolves to understand how their behavior changes in response to human visitation. Four motion-sensor cameras at each of 10 designated study sites throughout the park approximately 0.3 miles (0.5 km) away from one another at a 5-foot (1.5-m) height slanted downward to capture wildlife activity while maintaining privacy for human visitors by reducing the chance of capturing the faces on camera.[28]

The study's experimental design involved comparing wildlife activity in areas of human use with areas of low or no human use. During certain times, we designated study sites as tour vessel "heavy-use locations," where tour vessels were offered incentives to conduct shore excursions; during other times, the areas were off limits to shore excursions by tour vessels. This schedule was a key aspect of this project, giving us the ability to directly test the effects of human presence on wildlife activity. Sites were grouped in pairs: North/South Sandy Cove–Beartrack Cove, Reid Glacier West–Reid Glacier East, Adams Inlet–Hunter Cove, Lamplugh Glacier–Upper Tarr Inlet, and Bartlett Cove–Lester Island. Within each pair, 1 of the sites was designated for heavy use for the first half of the 2017 summer months (May through mid-July), and then the heavy-use designation switched to the other site for the following summer months (mid-July through September). We reversed the heavy-use schedule the following field season (summer 2018) to account for seasonal variation in wildlife activity. Bartlett Cove and Lamplugh Glacier remained heavy-use locations for the entire study, and several sites were not designated as heavy use.

Visual inspection of the detection data revealed 2 important thresholds: wildlife detections did not exceed 5 per week for any species unless human activity was absent, and wildlife detections at all backcountry sites dropped to 0 per week at human activity levels higher than approximately 30, roughly corresponding to 40 visitors per week. Patterns of wildlife activity in relation to human activity indicated that moose used human presence as a temporal shield and did not respond spatially, black bears exhibited a spatial response consistent with human shield effects while avoiding humans temporally,

As shore-based tourism continues to increase, it will be important to continue research on how human activity influences bears in the park. (NPS/Justin Smith)

wolves avoided humans temporally, and brown bears did not respond detectably to low numbers of humans in space or time. Lack of avoidance to low levels of human activity may indicate that brown bears may be habituated to people on the shoreline of the park, where the bears are protected from hunting and human activity is generally predictable. Authors of the study suggest limiting human use on beaches that exceed the threshold level of 40 visitors per week, with temporal zoning and concentrating human use in areas less desirable for wildlife.[29] The information provided by this study will help the park manage visitor use to provide for recreational opportunities while minimizing disturbance to wildlife and reducing the chance of significant resource degradation; it will also contribute to future backcountry and wilderness planning efforts.

FUTURE NEEDS

As shore-based tourism continues to grow in Glacier Bay, it becomes increasingly important to understand the effects of human activity on brown bears and other wildlife species. Future changes to the park's vessel quota may increase the number of vessels in Glacier Bay during the summer, which could subsequently increase brown bear disturbance and displacement. Park managers need a better understanding of how increases in the number of vessels and shoreline visitors will affect brown bear activity on the beach to allow for continued bear viewing in Glacier Bay while minimizing bear disturbance and displacement.

In the coming years, it will also be important to understand the effects of climate change on brown bears in Glacier Bay National Park and Preserve. Climate change will undoubtedly affect food availability and abundance, travel corridors, and the genetic population structure of brown bears in the park. Some impacts could be beneficial. Increased habitat as glaciers recede could allow for greater range expansion. Also, warmer weather can be good for flowers and berries, provided the plants receive enough rain during the growing season. However, detrimental effects from climate change may include diminished key food resources, particularly with loss of alpine habitat as tree level rises in elevation. Hot, dry weather can cause plants to green up early and dry out prematurely. Such conditions can also cause low creek levels and warm water when salmon are attempting to spawn, which greatly decreases their breeding success and abundance.

Ocean acidification will affect salmon, barnacles, mussels, and other important bear foods. Finally, as climate change melts glaciers that have kept brown bear populations separated for hundreds or thousands of years, as they encounter each other, genetic diversity may be lost due to genetic mixing. Studies focused on brown bear reliance on marine

Studies that focus on how bears use marine food resources will help us better understand the importance of these sources and impacts to them as the climate continues to warm. (NPS/Tania Lewis)

food resources across the park are needed to help managers understand and even forecast the effects of climate change on the distribution and abundance of these foods in the park.

One extreme example of how climate change may be detrimental to brown bear habitat in Glacier Bay National Park and Preserve involves glacial retreat of the Grand Plateau Glacier that feeds the Alsek River. Currently, the Alsek River flows out of Canada, through Alsek Lake, and into the Gulf of Alaska on the north side of the preserve. The river currently hosts abundant runs of multiple species of salmon accessible to bears in many different sloughs and tributaries between the lake and the mouth. Grand Plateau Glacier flows into Alsek Lake and is retreating rapidly, creating a new fjord. Geologists used laser altimetry and radar soundings to predict that as this retreat continues, the Alsek will abandon its present Dry Bay channel and flow out of Grand Plateau Lake instead, 17 miles to the south.[30] The area's topography is much steeper and the distance from the lake to the ocean is much shorter there. It is unlikely that bears will have the same salmon-foraging opportunities in the river's new channel. Understanding the full implications of this extreme landscape change on brown bear habitat and populations will help managers prepare for and predict the full effects of climate change on brown bears.

NOTES

1. Dauenhauer and Dauenhauer, eds. *Haa Shuká, Our Ancestors.*
2. Emmons, *The Tlingit Indians*; Newton and Moss, *The Subsistence Lifeway of Tlingit People*; Thornton, *Subsistence Use of Brown Bear in Southeast Alaska.*
3. Emmons, *The Tlingit Indians.*
4. Dauenhauer and Dauenhauer, *Haa Shuká*; Swanton, *Tlingit Myths and Texts.*
5. Swanton, *Tlingit Myths and Texts.*
6. White, interview with Mary Beth Moss.
7. Milner et al., "Colonization and Development of Stream Communities across a 200-Year Gradient in Glacier Bay National Park"; Howe, *Bear Man of Admiralty Island*, 61–83.
8. Visitor statistics for all parks can be found at https://irma.nps.gov/Stats/Reports/Park/GLBA.
9. National Park Service, *Glacier Bay Bear-Human Management Plan.*
10. Lewis, Stanek, and Young, *Bears in Glacier Bay National Park and Preserve.*
11. Lewis, Stanek, and Young, *Bears in Glacier Bay National Park and Preserve.*
12. Young and Lewis, *Bears of Glacier Bay National Park and Preserve*; Partridge, Smith, and Lewis, *Black and Brown Bear Activity at Selected Coastal Sites in Glacier Bay National Park and Preserve.*
13. Lewis, Stanek, and Young, *Bears in Glacier Bay National Park and Preserve.*
14. Lewis, Stanek, and Young, *Bears in Glacier Bay National Park and Preserve*; Young and Lewis, *Bears of Glacier Bay National Park and Preserve.*
15. Lewis, Stanek, and Young, *Bears in Glacier Bay National Park and Preserve.*
16. Lewis, Stanek, and Young, *Bears in Glacier Bay National Park and Preserve.*
17. Lewis, "Shoreline Distribution and Landscape Genetics of Bears."
18. Lewis, "Shoreline Distribution and Landscape Genetics of Bears."
19. Lewis, Stanek, and Young, *Bears in Glacier Bay National Park and Preserve.*

20. Pinjuv, "Estimating Black Bear Population Size in Gustavus, Alaska."

21. Pinjuv, "Estimating Black Bear Population Size in Gustavus, Alaska."

22. Mattson, Herrero, and Merrill, "Are Black Bears a Factor in the Restoration of North American Grizzly Bear Populations."

23. Pinjuv, "Estimating Black Bear Population Size in Gustavus, Alaska."

24. Lewis, Pyare, and Hundertmark, "Contemporary Genetic Structure of Brown Bears (*Ursus arctos*) in a Recently Deglaciated Landscape."

25. Lewis and Lafferty, "Brown Bears and Wolves Scavenge Humpback Whale Carcass in Alaska."

26. Sytsma et al., "Low Levels of Outdoor Recreation Alter Wildlife Behavior."

27. Lewis, Stanek, and Young, *Bears in Glacier Bay National Park and Preserve*.

28. Sytsma et al., "Low Levels of Outdoor Recreation Alter Wildlife Behavior."

29. Loso et al., "Quo vadis, Alsek."

30. Loso et al., "Quo vadis, Alsek."

A young bear stands to get a better look into the river to spot fish. (NPS/Lian Law)

Brown bear research in Denali (at the time called Mount McKinley National Park) was pioneered by the famous naturalist Adolph Murie in the late 1950s and into the 1970s. His long-term observations were valuable in describing brown bear ecology and were the foundation for research today. (NPS/Daniel A. Leifheit)

8 DENALI

FROM MURIE TO MODERN DAY

Patricia Owen

HISTORY

Athabascan people have made their homes on lands within what is now Denali National Park and Preserve for perhaps as long as 12,000 years. Their relationship with wildlife was primarily related to subsistence, focused on moose and caribou. Some area inhabitants harvested black bears for food; however, cultural beliefs about bears often limited the preparation and consumption of bear meat to men and older women. Therefore, bear meat was not likely a substantial part of their diet. Bear grease rendered from the fat was eaten and believed to have had medicinal properties when used topically. Bears were believed to be highly intelligent and deserving of respect. Women of childbearing age avoided any contact with bear parts to assure the safety of their unborn children. Prior to a hunt, bears were not called by name but referred to as "it" in fear that the bear might learn the hunters' plans. If the hunt was successful, the bear received an apology for mistakenly being killed and was also thanked for giving its life. Often, the bear's eyes were removed so it could not see the hunter. The tendon under the tongue was cut to keep the bear from warning

"Wild grizzlies on Mt. McKinley National Park, conducting their affairs undisturbed, are the essence of wilderness spirit."
—ADOLPH MURIE

https://doi.org/10.5876/9781646427116.c008

other bears. After butchering, the bear's spirit was released by leaving its head on the top of a tree. Hunting bears was generally preferred in the fall, when they were found near salmon streams, and in early winter, when they were killed in their dens.[1]

Hunter-naturalist Charles Sheldon was one of the first Europeans to study the wildlife in the Denali region. From 1906 through 1908, as Sheldon hiked, he hunted, collected, and observed a range of mammals. The foci of his pursuits were the "white sheep" (Dall's sheep), but brown bears also featured prominently in his observations and collections. He described bears hunting mice in the snow and ground squirrels in their burrows and remarked on the absence of berries in the stomachs of bears he collected, even in areas where berries were abundant. Sheldon's appreciation of the grandeur of the mountains and the wildlife contained therein prompted him to lobby the United States Congress to protect the area, leading to the establishment of Mount McKinley National Park in 1917.[2]

The land was originally set aside for the protection of its wildlife populations, not because of its physical landscape or the tallest peak in North America, Denali, formerly known as Mount McKinley. At that time, no roads connected Denali to the rest of the state. Most visitors arrived by train.

Joseph Dixon made numerous trips to what was then Mount McKinley National Park in the 1930s. Dixon documented that brown bears typically do not kill adult ungulates but do take advantage of winter-killed carcasses in early spring. He also emphasized the importance of rodents (like ground squirrels) in brown bears' diet. Interestingly, like Sheldon, he reported that even when abundant, brown bears did not eat berries to a great extent.[3]

Famed naturalist Adolph Murie conducted research in the park for decades. His pioneering research on brown bears was the result of intensive field observations mainly between the late 1950s and the 1970s. Murie described the behavior of lone bears as well as that of mothers and cubs and how they behaved around other bears. He studied home range, diet, and brown bears' relationships with other animals. His observations of brown bears for multiple summers allowed Murie to make inferences that other researchers, who had only observed bears for a few weeks, could not. Thus, he was able to use what he observed to paint a broader picture of bear ecology. He found that home ranges overlap but territoriality was not an issue.

By observing breeding pairs, Murie determined that the breeding season for brown bears extended from mid-May to early June. He found that brown bear cubs remain with their mother until age 2½, during which time they learn survival skills through imitation, play, and practice. Bears are omnivorous, but they rely on vegetation for the bulk of their diet. Cubs learn to dig roots and ground squirrels, preferred grasses, herbs, and berries, as

A 1939 photo taken by Adolph Murie of a brown bear at Igloo Creek in Mount McKinley National Park. (NPS Historic Photograph Collection, HFCA 1607; NPS/Adolph Murie)

well as how to catch young ungulates and defend carcasses against other bears and other predators. During his observations, Murie documented interactions between brown bears and other animals, particularly wolves. Brown bears tend to benefit from this relationship because they feed on kills made by wolves—prey that would be difficult for bears to obtain on their own. The results of his work were published posthumously in a volume titled *The Grizzlies of Mount McKinley*.[4]

Changes in Denali were afoot. Despite's Murie's efforts, the park road was extended and now reaches to Kantishna. By the late 1950s, the Denali Highway was complete, connecting Paxson and the Richardson Highway to Cantwell, and park visitation began to rise. Open pits for trash disposal and bear viewing at these locations were a popular attraction. In 1972, the George Parks Highway was completed, and visitation took another leap. Few camping guidelines were in effect, no wildlife safety education was provided for visitors, and bear-resistant food storage facilities were poor to nonexistent. Park managers realized that food conditioning was bad for bears and caused conflicts with people, so open trash dumps were closed in the mid-1970s. In 1978, the first bear-human conflict management plan was developed. It included the creation and use of a bear incident form, some bear-safety education, and the installation of "bear-proof" trash cans.

The Alaska National Interest Lands Conservation Act (ANILCA) in 1980 more than doubled the size of the park and established adjoining preserve areas. The new law also renamed the park Denali National Park and Preserve. Conservation of brown bears and their habitats was explicitly noted as a fundamental purpose of the park. Unfortunately, by 1982, Denali had the highest rate of backcountry human-bear incidents of any U.S. national park, with high backcountry use and a significant bear population.

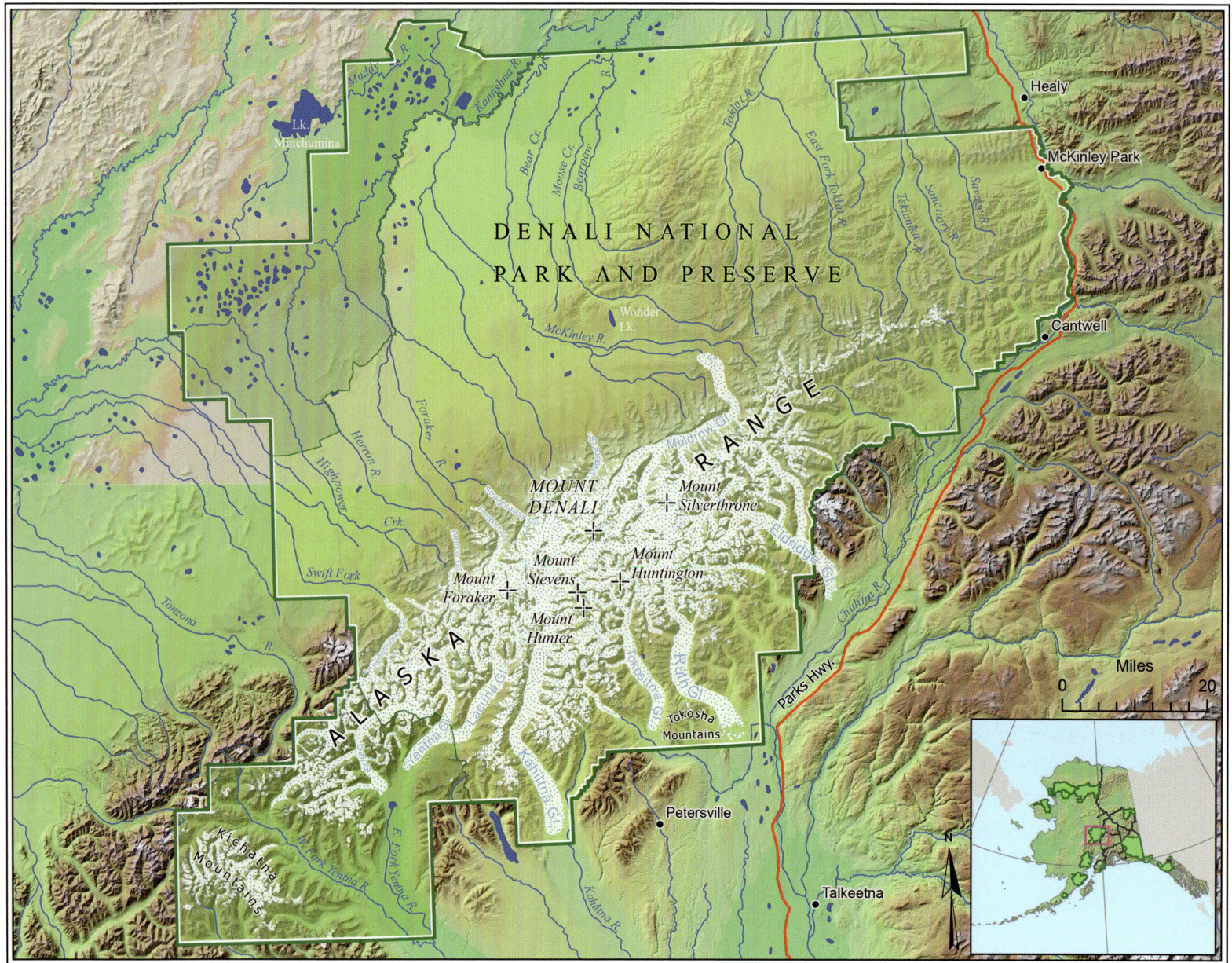

The spine of the Alaska Range bisects Denali National Park and Preserve, with the summit of Denali reaching over 20,000 feet.
This creates distinct north and south climatic regions and biological diversity in the park.

A new management plan was developed, and bear-resistant food containers were tested and encouraged for backcountry campers. The first "bear technician" was hired to implement the plan. Francis Singer, along with Joan Beattie, noted that human-bear incidents increased with increased visitation and found that bear sightings declined as visitation increased.[5] John Dalle-Molle conducted some of the first tests of bear-resistant food containers for use by backpackers and determined that they were successful at keeping bears away from human food. He documented a 74% reduction in bears acquiring human food.[6] In addition, he found that backcountry hikers were mostly positive about using the containers. In 1987, bears did not acquire any food from backpackers, and since that time the park has had only 1 human fatality from a bear mauling, which occurred in 2012. Dalle-Molle and Joseph Van Horn revamped the park's human-bear conflict management program and created a human-bear conflict management plan that was a model for all subsequent bear plans. They were instrumental in the success of this emerging program and provided the groundwork for similar bear management plans around the National Park Service. The basic tenets of their work still support bear management in Denali today.[7]

ENVIRONMENT

The park ranges in elevation from just over 200 feet in some of the lowlands in the southwest preserve to over 20,000 feet at the summit of Denali. The spine of the Alaska Range roughly bisects the park from east to west and creates 2 distinct climates in the park. Temperatures north of the Alaska Range are typical of an interior Alaska climate, with very warm summers and cold winters. At park headquarters, temperature extremes range from 91°F (33°C) to –54°F (–48°C). The average high temperature in July is 66°F (19°C), and the average low temperature in January is –6°F (–21°C). Total annual precipitation is relatively low, at 16 inches (406 mm); more than half of that falls as rain in the summer. Snow covers the ground from October through May. The seasonal snowfall total, on average, is only about 77 inches (196 cm), but it persists because of the cold. The transitional maritime climate on the south side of the Alaska Range is influenced by the prevailing weather patterns of the Gulf of Alaska, with milder air temperatures, less seasonal variation, and more precipitation. Along the southern flanks of the Alaska Range, snowfall is abundant, and snow cover is often present through late spring.

The distinct north and south climatic regions result in different biological characteristics in the park. Vegetation in the park is highly diverse and includes alpine tundra, shrub-scrub tundra, mixed spruce-birch and spruce-tamarack forests, taiga, wetlands, and extensive riparian and lowland forests. The southern area is vegetated by extensive expanses of white spruce forests interspersed with wide, sweeping glaciers. Rivers originating from these glaciers flow south and eventually into Cook Inlet. Glaciers on the north side of the Alaska Range are the origins of rivers that make their way to the Yukon River farther northeast. The park contains over 10,000 mapped lakes and more than 753 species of flowering plants.

Both brown bears and American black bears reside in the park and preserve. Brown bears, however, are the dominant focus of most bear research and management activities, for several reasons. Brown bears tend to be found in relatively open tundra where they are easily observed, they are of great interest to park visitors, and they tend to interact with those visitors more often than black bears do. Adolph Murie did focus some of his field observations on black bears, and black bears have sometimes been studied incidental to other bear projects.

MANAGEMENT CHALLENGES

The biggest management challenge related to brown bears in Denali is human-bear interactions. Access to much of the park and preserve is challenging, so the bulk of park users enter the park by road at the east boundary and travel the Denali Park

Road, which extends 90 miles to the west where it ends in Kantishna. With visitation topping 600,000 visitors each year and an estimated 300–350 brown bears, observations of bears and human-bear interactions are common. The current bear management program relies heavily on bear-safety education for all park users in addition to bear-resistant food and trash containment and bear-viewing distances to keep bear and people safely separated. Increasing visitation and the infrastructure to support it will continue to put pressure on bears and their habitat.

Through the 1970s and 1980s, many people conducted research related to the effects on bears of people as well as of traffic along the Denali Park Road. Frederick Dean studied brown bear population ecology and looked at density, age, and sex composition and status; litter size relationships; and mating habits. He conducted some of the first aerial brown bear density estimates for the park. He was joined by others who reported on reactions to human activity along the Denali Park Road, bears at garbage dumps in Denali, interactions between backpackers and brown bears, bears killing other bears, mixed-age bear family groups, and brown bear habitat use. These issues remain relevant today.[8]

The other primary management issue deals with brown bear harvest. The 1980 expansion of Denali separated the area into 3 different zones, each with slightly different mandates governing their use. Wildlife-related regulations in the 3 areas focus

Human-bear interactions are the biggest management challenge in Denali National Park and Preserve. Although most people see bears from the safety of a tour bus, with more than 600,000 visitors and approximately 300–350 bears in the park, interactions are common. (NPS/Claire Abendroth)

on harvest: hunting and trapping. No harvest of wildlife can occur in the old Mount McKinley Park, all of which is now federally designated wilderness. Subsistence harvest occurs in the new park and preserve additions, and sport hunting is allowed in the preserve. Wildlife harvest in the preserve is allowed and regulated by Alaska state game laws, as long as they are compatible with NPS mandates. Changes in state regulations to allow for longer seasons, increased bag limits, and additional means and methods make bears crossing park boundaries more vulnerable to harvest outside park boundaries.

RESEARCH SUMMARY

Many people have contributed to our understanding of bears in Denali over the years. Countless hours have been spent observing bears living out their lives as the impact of human presence continually increases around them. The earliest researchers depended on foot travel and a good pair of binoculars. As technology advanced, the methods changed. Motor vehicles on the park road made travel faster, and aircraft allowed bears to be studied in some of the most remote parts of the park. Radio collars made it possible to positively identify the same bear year after year. With the advent of satellite radio collars, researchers could follow

Human-bear interactions can result in a need to relocate or haze the bear away from areas of high visitor use. (NPS/Dave Schirokauer)

bears without leaving the office. The early 1990s witnessed the start of a long-running series of bear research studies in Denali that are ongoing. Though the objectives varied, all of these projects relied on radio telemetry.

Brown Bear Population Ecology and Monitoring, 1991–1998

Jeffrey Keay studied the dynamics and ecology of the naturally regulated brown bear population in Denali. This study was pivotal in determining vital rates for Denali brown bears. Bears were captured

◁ Wildlife biologist Pat Owen and pilot Troy Cambier collar a bear in Denali. The radio collar studies provided important information about bear emergence in the spring and how they use the park landscape. (NPS)

Cubs of the year experience the lowest survival rate, with only approximately 37% surviving to den at the end of the year. Survival increases with age, with a 79% survival rate for 2-year-old cubs. (NPS/Jacob Frank)

by helicopter and fitted with radio collars. They were located at least monthly during the active period (about April through October), with particular focus on the periods of den emergence in spring and den entrance in fall. Survival rates were estimated by age class. Cubs of the year experienced the lowest annual survival rate, at 0.371, meaning that nearly 37% of all cubs born in any year survived to go to den in fall. Survival of cubs increased with age, with a rate of 46% for yearlings and 79% for 2-year-olds. Adult survival jumped to 97% for females and 98% for males. Females were found to have their first litter of cubs between ages 6 and 10. Cubs became independent of their mothers at about 3 years. Bears were screened for diseases including canine hepatitis, canine distemper, and leptospirosis, but low antibody prevalence indicated that none were important factors in population dynamics. During the fall of 1995, the density of brown bears was estimated using capture-mark-resight methodology, with an estimated 61% of the population marked. Density was estimated at 27.1 independent bears/100 km². Keay determined that the high survival rates of independent bears and lack of human interference suggest that the Denali brown bear population is likely at carrying capacity.[9]

Resource Selection of Brown Bears and Black Bears in South-Central Alaska, 1998–2000

As a result of increased demand for visitor access to Denali, the National Park Service led a multi-agency planning effort to develop visitor facilities for portions of Denali south of the Alaska Range, Denali State Park, and adjacent areas. The plan included construction of visitor centers with associated campgrounds, primitive fly-in campsites, public-use cabins, and hiking trails. It was anticipated that this development would increase south-side visitor use from about 20,000 to 250,000 annually.

A major concern was balancing the demands for increased recreational activities with the need for increased protection of critical wildlife habitats. It was essential that managers have access to high-quality scientific information to formulate and justify decisions and monitor human-induced effects on the environment.

Brown and black bears are important components of the wildlife-viewing experience for many visitors to Denali National Park and adjacent Denali State Park. These species are also important in providing opportunities for hunting in some areas. In addition, the parks have developed and maintained human-bear conflict management programs. The continued success of these programs depends, in part, on a thorough understanding of bear ecology and techniques to resolve human-bear conflict.

Current ecological data on bears in Denali at the time had focused on bears in tundra-dominated habitat on the north side of the Alaska Range. Climatological differences on the south side have resulted in forest-dominated habitats unique to those found in the remainder of Denali. In addition, bears' prey base varies between the 2 areas. For example, salmon are abundant on the south side and more limited to the north. As such, current habitat data available for bears from Denali are not applicable to land-use planning on the south side.

The overall goal of a project conducted by Jerrold Belant was to estimate the effects of proposed human development on brown and black bear resource selection—specifically, determining seasonal habitat use of brown and black bears. He found that overall, brown bears used the shrub cover type (38%), followed by deciduous forest (25%) and tundra (18%). Overall, black bear habitat use was similar, with shrub (36%), deciduous forest (31%), and tundra (12%) used most frequently. Spruce forest was used the least by each species (≤5%).[10]

Since salmon are sought by both bears and humans, it was important to determine bears' proximity to streams when salmon are present. Overall, brown bears were closer to streams than were black bears. Mean distances of brown bears from streams were lowest during late June through July and mid-August through early September, which coincides approximately with salmon runs.

Belant described brown bear and black bear resource selection in the southern reaches of Denali, the way those resources are partitioned between the species, and how that partitioning affects diet and reproduction. Based on isotopic analysis, salmon comprised more than 53% of the diet of brown bears, whereas salmon made up no more than 25% of black bear diets; black bears avoided areas occupied by brown bears in the summer. Limited access to salmon forced black bears to forage in areas with less nutritious food, resulting in suboptimal body condition and thus lower reproduction.[11]

Habitat Use and Movement of Brown Bears in Denali National Park Relative to the Denali Park Road, 2006

Richard Mace and others worked on an integrated study of the capacity of the Denali Park Road that was designed to ascertain whether the road impacted the habitat use, behavior, and movements of brown bears. Many animals respond negatively to human activity along roads, and roads may present barriers to movement.[12]

◄ Bears use shrub, forest, and tundra habitats throughout the park. (Stuart Leidner)

In this study, the movements and activity patterns of 17 brown bears were analyzed. Most (13) were female bears. Eleven of these bears had home ranges that straddled the road, 4 abutted the road, and 2 bear ranges were considered far from the road. Movements of these study bears spanned the length of the Denali Park Road. Brown bears in this study were most active during the period of day when road traffic was most pronounced, which coincided with daylight hours. As evidenced by other studies, this pattern of relatively high activity during the daylight hours is the norm for brown bears across their range. The fact that these brown bears were most active during periods of high traffic suggests that they were not altering their temporal patterns of activity to avoid human presence along the road.[13]

Periods of inactivity for brown bears were more confined to hours of darkness. Brown bears in Denali exhibited periods of relative inactivity up to 14 hours in duration. For both early and late seasons, periods of inactivity (movements of <10 m per hour) were most likely to occur during hours of darkness. There was a relationship between the duration of inactivity and distance to the Denali Park Road. Periods of inactivity were shortest when bears were nearest to the road and increased in duration as distance to the road increased. These data suggest that bears were less comfortable being either relatively stationary or sleeping near the road corridor.[14]

Eleven brown bears were documented crossing the road a total of 444 times during this study. Both male and female brown bears crossed the road at all hours of the day, in both the early and late seasons. Peak times of crossing the road occurred when traffic levels were moderate to high. The speed at which brown bears moved increased significantly as they moved across the road. The increase in movement speed while crossing suggests that bears were cognizant of human activity along the road and used speed to minimize the duration of contact with humans or vehicles.[15]

Trans-Boundary Movements in Northeast Denali, 2010–2018

Bears in and around Denali are high value for both viewing and harvest. Bear-viewing opportunities are typically abundant from the bus system along Denali Park Road and in the backcountry. Harvest of bears is not allowed within the original park, but subsistence harvest is allowed in the new park and preserve additions, and sport hunting is allowed in Denali National Preserve. Much harvest occurs just outside the northeast corner of the park near areas that receive high visitation. Bear baiting is a harvest technique recently allowed under State of Alaska sport hunting regulations, and the area is designated an Intensive Management Area, meaning that predator control (intentionally killing bears to reduce their numbers) could be implemented. A question that became apparent is whether "resident" park bears move across the park boundary into unprotected areas and at times that would make them vulnerable to harvest. If they do and are removed by harvest, would viewing opportunities within the park be affected?

To answer these questions, I deployed GPS (Global Positioning System) collars on 40 brown bears between 2010 and 2018 (13 male, 27 female). Auto-correlated kernel density was used to calculate home range size. The average home range size was 365 square miles (945 km²). Twenty-one of 40 bears (53%) went outside the park at least once during monitoring. Based on average home range size, bear viewing on about half of the park road could be affected by harvest. Of bears that traveled outside the park, 13 made extensive forays. Thus, 33% of bears spent considerable time outside the park. Many forays were to salmon streams, and most occurred during harvest season.

In conclusion, bears in Denali have large home ranges typical for this ecosystem. Although no study bears were definitively harvested, bears visible along a large portion of the park road may be vulnerable to harvest.

Road Traffic Impacts, 2022–2026

The influence of the park road on brown bears and their movements is not completely understood to this day. Near the end of the summer season in 2021, park managers made the decision to close the Denali Park Road. A landslide at Polychrome Pass had exceeded the efforts of the road maintenance crew, and the road had become impassable. It quickly became evident that the road would require extensive construction that would take at least 2 summer seasons to complete. This closure effectively means that the western half of the road will be free of normal traffic during this time, thus providing an unexpected opportunity for a natural experiment to study bears' movements and behavior in a "no-traffic" condition. Comparison to a more "normal" traffic condition will be possible after the road is repaired and normal traffic resumes. The study began in 2022, when staff opportunistically recorded bear observations and behavior along the closed section of road. In 2023, a sample of bears was fitted with GPS collars to determine movements in relation to the road corridor when little to no traffic is present. After collecting location information for 2 active seasons, the same bears will be fitted with fresh collars to coincide with resumption of a "traffic" condition on the road. In this way, a direct comparison of road corridor use by bears without and with traffic present will be possible.

CAUTION
HEAVY
EQUIPMENT
WORKING
AHEAD

Please Establish
Eye Contact With
Operator Before
Passing

NOTES

1. Stokes, *Natural Resource Utilization of Four Upper Kuskokwim Communities*.
2. Sheldon, *The Wilderness of Denali*.
3. Dixon, *Birds and Mammals of Mount McKinley National Park, AK* cf. p. 229.
4. Murie, *The Grizzlies of Mount McKinley*.
5. Singer and Beattie, "The Controlled Traffic System and Associated Wildlife Responses in Denali National Park."
6. Dalle-Molle, "Field Tests and Users' Opinions of Bear Resistant Backpack Food Containers in Denali National Park."
7. Dalle-Molle and Van Horn, "Bear-People Conflict Management in Denali National Park."
8. Dean, "Brown Bear Density, Denali National Park, Alaska."
9. Keay, *Grizzly Bear Population Ecology and Monitoring, Denali National Park and Preserve, Alaska*.
10. Belant, *Resource Selection of Brown Bears and Black Bears in Southcentral Alaska*.
11. Belant, "Resource Partitioning by Sympatric Brown and American Black Bears."
12. Mace et al., *Habitat Use and Movement Patterns of Grizzly Bears in Denali National Park*.
13. Mace et al., *Habitat Use and Movement Patterns of Grizzly Bears in Denali National Park*.
14. Mace et al., *Habitat Use and Movement Patterns of Grizzly Bears in Denali National Park*.
15. Mace et al., *Habitat Use and Movement Patterns of Grizzly Bears in Denali National Park*.

◀ Bears are naturally curious and investigate their environment. As the Denali Park Road is closed and construction work is completed, it will be interesting to see if or how bears' movements through the park change. (NPS)

9 GATES OF THE ARCTIC

BEARS IN THE BROOKS RANGE

Kyle Joly, Mathew S. Sorum, Matthew D. Cameron, William Deacy,
Grant V. Hilderbrand, Susan Georgette, and David D. Gustine

HISTORY

Brown bears and people have coexisted in what would become Gates of the Arctic National Park and Preserve for more than 10,000 years. Brown bears have long held a revered place in the traditional beliefs, practices, and ways of life of the Iñupiat and the Koyukon Athabascans, Alaska's Indigenous peoples with homelands in the central Brooks Range. Both of these Indigenous cultures have lived alongside and closely observed bears for untold generations. Both also have tremendous respect for these animals, attributing to them special spiritual as well as physical powers, and they have rules for personal behavior that, if broken, may doom a hunt or the hunter.[1]

Central to these rules of appropriate human behavior toward brown bears is an attitude of humility and deference. Elders instruct hunters not to either brag about how many bears they have caught or speak about bears in a confident or threatening manner. Hunters should not talk about their intentions to hunt these animals.[2] As a sign of respect, both the Iñupiat and the Koyukon avoid saying their term for bear,

◄ Gates of the Arctic National Park and Preserve is entirely above the Arctic Circle. It includes the central Brooks Range, the northernmost extension of the Rocky Mountains. (NPS/Kyle Joly)

https://doi.org/10.5876/9781646427116.c009

preferring to name these animals obliquely with Iñupiaq or Koyukon words that translate to "one who walks," "those who are in the mountains," "big animal," or a similarly respectful term.

Iñupiaq and Koyukon hunters must follow certain practices after a successful hunt to show the bear proper respect, the details of which vary among the 2 groups. Among the Koyukon, some prescribed behaviors center around women and bears, which have ties to an origin story in which a bear becomes a woman's husband.[3] Similarly, the Iñupiat have origin stories of bears that feature humans, implying a close relationship between people and bears.

Gates of the Arctic was indirectly named by Bob Marshall, pioneering wilderness explorer and advocate.[4] During his travels along the North Fork of the Koyukuk River around 1930, he came to the location where the Boreal Mountains and Frigid Crags penned in the river and saw those peaks as towering gates through which to enter the Arctic region to the north. His descriptive writing was so captivating and enduring that the park was eventually named for the description he gave the peaks. Marshall ranged widely out of the village of Wiseman, exploring the mountains of the Brooks Range. During these adventures, he came across brown bears. One day, while exploring Amawk Creek, he came within 150 feet (46 m) of "three grizzlies"—presumably a sow and 2 older cubs—that stood on their hind legs to check him out and then disappeared into the willows.[5] "Grizzly" bears or "grizzlies" are commonly used names for brown bears in interior (inland from the coast) North America. The name is thought to have originated from the Lewis and Clark Expedition (1803–1806), when the expedition members became the first westerners to provide detailed documentation of the "grisley" bear.[6] Whether the name stems from its "grizzled" (gray) streaked hair or its supposedly "grisly" behavior (horrifying or gruesome) is not entirely clear. However, the latter may have helped inspire the scientific name, *Ursus arctos horribilis*. Marshall named a tributary of the North Fork of the Koyukuk River, just a few miles from Amawk Creek where he had his heart-stopping encounter, Grizzly Creek.

Gates of the Arctic National Park and Preserve was established on December 2, 1980, by the Alaska National Interest Lands Conservation Act (ANILCA). The park's roughly 8.5 million acres (3.5 million ha) were set aside to maintain the wild character of this vast landscape. The protection of habitats for and populations of "grizzly bears" is explicitly mentioned as a fundamental purpose of the park.

▶ A brown bear trekking across tundra in fall colors. (NPS/Erika Jostad)

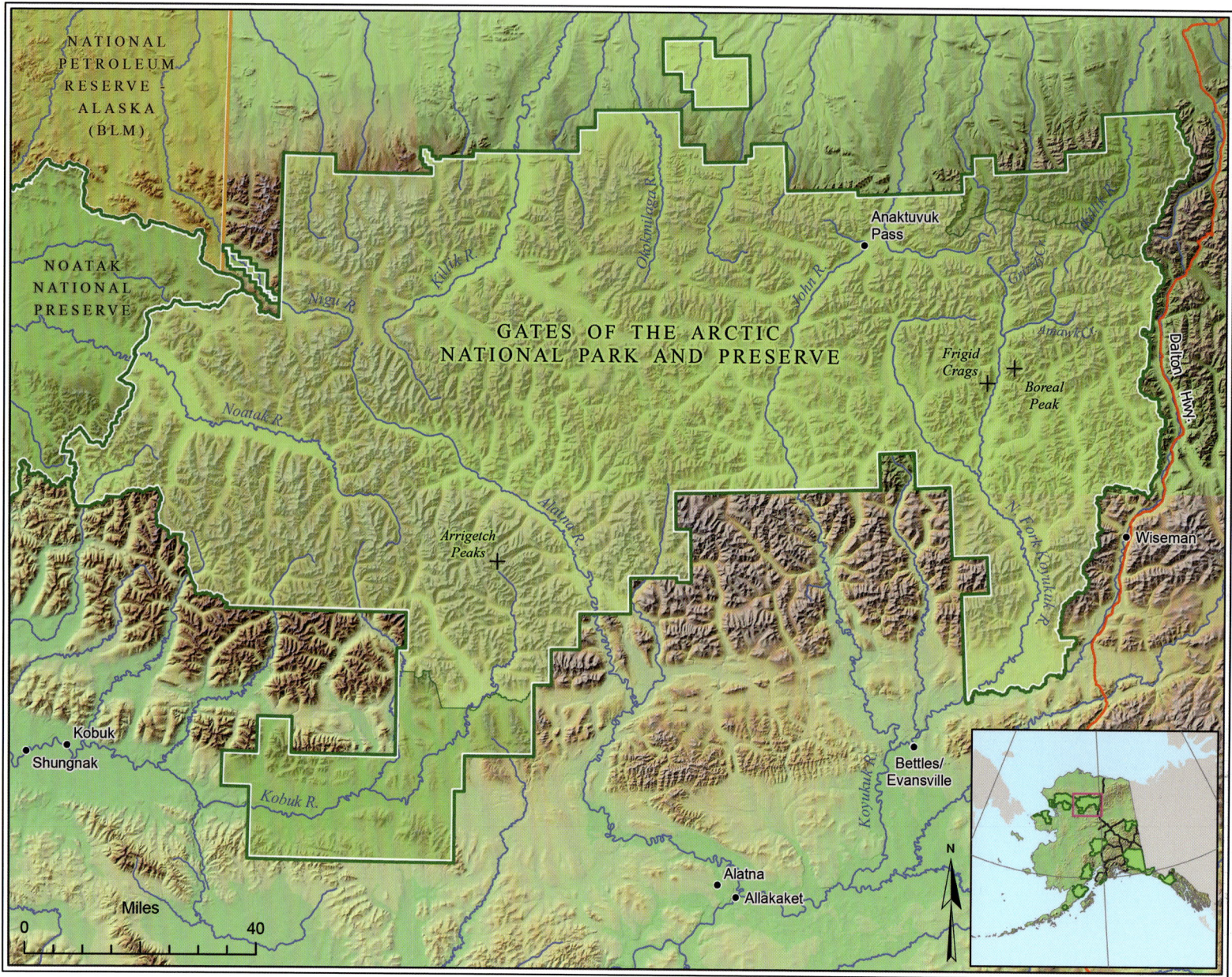

Gates of the Arctic National Park and Preserve was established in 1980, with an explicit purpose to protect the habitat and populations of grizzly bears.

Pioneering work on brown bears in northern Alaska began in earnest in the 1980s. Biologists including Warren Ballard, Dick Shideler, Harry Reynolds, Gerald Garner, John Hechtal, and others worked in the Brooks Range and Alaska's North Slope. Much of this work was conducted to the north, east, and west of Gates of the Arctic National Park and Preserve. We have been unable to discover any scientific publications on brown bears that focused on Gates of the Arctic throughout the history of the park until our work began.

ENVIRONMENT

Gates of the Arctic National Park and Preserve covers the central Brooks Range—the northernmost extension of the Rocky Mountains and the Continental Divide—and is entirely above the Arctic Circle. These rugged, largely barren mountains reach up to about 8,500 feet (2,600 m) in height. Thickets of alder, willow, and dwarf birch shrubs cover the middle slopes but give way to the boreal forest at lower elevations on the southern slopes of the range. Typical of Arctic environments, vegetative productivity is low. A 100-year-old black spruce tree may be less than 9 feet (3 m) tall. In the central and northern portions of the park, trees are absent due to extreme cold, and Arctic tundra dominates the lower elevations.

In the Arctic, temperatures can drop to –50°F (–46°C) in winter, and annual average temperatures are well below freezing, at 16°F (–9°C). Across the Brooks Range, winter days are short, and the sun does not rise on the solstice. Snow can reach 50 inches (127 cm) in depth. Winter conditions, with below-freezing temperatures and snow on the ground, can persist for 8 months (October through May). However, the Arctic is a land of extremes, and temperatures on the south side of the Brooks Range can reach 90°F (32°C) during its brief summers. With these high temperatures and 24 hours of daylight, wildfires are common in the boreal forest in the southern part of the park. Large berry crops, including blueberries and crowberries, occur during cooler, wetter summers.

Waters streaming out of Gates of the Arctic National Park and Preserve flow in 3 directions. In the southeastern portion of the park, tributaries, such as the Alatna River and the John River, feed the Koyukuk River, which eventually flows into the Yukon River and the Bering Sea. The western waters flow west into the Kobuk and Noatak Rivers, which flow directly into the Chukchi Sea. In the northern portion of the park, the waters flow north from tributaries to the Colville River and eventually reach the Arctic Ocean. Arctic grayling, whitefish, and Arctic char live in these waters, and salmon and sheefish are known to inhabit the Kobuk River system.

While brown bear productivity is typically low in Gates of the Arctic National Park and Preserve, some bears, like this sow with three cubs, are more productive than others. (USFWS/Dan Nolfi)

Salmon were known to spawn in lower stretches of the Noatak and Koyukuk River systems, but there was little to no documentation of their presence within the park until our research began.

Gates of the Arctic National Park and Preserve is home to a complete set of native wildlife species. Many of these species, including moose, caribou, Dall's sheep, and Arctic ground squirrels, are prey for brown bears. Dall's sheep tend to stay close to very rugged terrain during the summer months when bears are active, and they can usually evade predation. Caribou are typically present only in the northwestern portion of the park during bears' active season. Extremely fleet of foot, even 2- to 3-month-old caribou calves can typically outrun bears. The calving grounds for caribou are far away from Gates of the Arctic, so younger calves are not available for bears in the park. Moose calve in mid- to late May along the southern flanks of the Brooks Range, and the newborns are available for brown bears to catch, but the greatest impact lasts only a few weeks before the calves are large enough to effectively escape predation. It is also during this time that bears may discover and eat large mammals killed by winter starvation or avalanches. Low vegetative productivity in the Arctic translates into relatively low productivity of prey species of brown bears. Given the low densities of large prey and the limited time they are available to bears, it was assumed that many bears in the park rely heavily on vegetation in their diet. Here, brown bears inhabit an extreme environment, with their primary foods limited in space and time. These extremes and limitations are why Gates of the Arctic is near the northern range extent for this species.

MANAGEMENT CHALLENGES

Perhaps the biggest management challenge facing Gates of the Arctic National Park and Preserve with respect to brown bears was a complete lack of data prior to recent research (see the next section, Research Summary). From its founding in 1980 until 2010, the park did not undertake any studies or conduct any surveys of brown bears. In late May and early June 2010, park biologists estimated the number of brown bears in the northeastern portion of the park. While providing a baseline level on the number of brown bears in the park, the estimate shed little light on the ecology of these bears living at the northern fringe of their range.

A growing management challenge is increasing visitor use. As with the National Park System in general, visitation to Gates of the Arctic is increasing. The number of bear-human conflicts correlates directly with the number of people using an area. Because the park still has relatively low visitation, it has had relatively few human-bear incidents. Indeed, it is possible that some bears in the park have never seen a human.

We know of only 1 death attributed to brown bears in Gates of the Arctic National Park and Preserve since its creation in 1980. In 1996, 2 hikers walked into a willow thicket along a tributary of the Noatak River and surprised a bear, which fatally injured 1 of them. The only other severe mauling was in 2008, when a woman camping along the Okokmilaga River was dragged out of her tent by a bear. The vast majority of bear encounters occur in locations with relatively high human visitation, such as the Arrigetch Peaks area. Gates of the Arctic, like all other national parks, is tasked with protecting its resources—such as brown bears and wilderness—while also allowing for visitor enjoyment. Thus, educating people who visit the park about bear ecology and behavior, as well as how to stay safe in bear country (and the entire park is bear country), is critical to bear management and especially for reducing human-bear incidents. Keeping food and other attractants away from bears is one of the most important steps in reducing problems; thus, the park requires visitors to use "bear barrels" (bear-proof containers to store bear attractants such as food and toothpaste) while camping in the park. When park management priorities are in conflict, resource protection overrides visitor use.[7] For example, the number of people and the period of time when they can use an area can be limited to improve resource protection. Gates of the Arctic is largely a wilderness park and does not have infrastructure like campgrounds and roads.

While it was not much of a threat to the park and its bears during the first 40 years of its existence, development is now a critical concern. Petroleum development has vastly increased on Alaska's North Slope since Gates of the Arctic was established, especially within the National Petroleum Reserve–Alaska (NPR-A) to the north and northwest of the park. With these developments have come proposals for additional developments, like industrial roads connecting North Slope communities as envisioned by the State of Alaska's Arctic Strategic Transportation and Resources Plan. Another, the "Roads to Resources" industrial road from the Dalton Highway to the Ambler Mining District, known as the Ambler Road, was temporarily permitted in 2020; the selected route runs through the Kobuk River section of the park. Discussion about this proposed road increased in 2012 and was the impetus for launching a brown bear ecology research project in 2014. As of this writing, the permit is the subject of ongoing debate.

RESEARCH SUMMARY

To address our massive information gaps, especially relating to the potential development of the Ambler Road, we conducted an ecological study of brown bears from May 2014 to September 2017. We deployed GPS (Global Positioning System) collars on adult female and male brown bears. Additional captures and the recapture of previously collared individuals occurred in June 2015 and June 2016. A total of 33 females and 19 males were collared during the study. At the time of capture, in addition to collaring individuals, the bears were sexed, aged, and weighed; their size was measured; and the number of offspring was counted. Also, we collected blood and hair samples for assessment of diet, pathogens, genetics, and contaminants.

We found that interior brown bears in Gates of the Arctic National Park and Preserve are much smaller than their counterparts in coastal parks, like Katmai National Park and Preserve. Half of the adult females weighed less than 200 pounds (91 kg), and the heaviest weighed just 340 pounds (154 kg). Adult male bears were larger than females and showed greater variability in weight, ranging from 146 pounds to 553 pounds (66–251 kg). Both male and female bears in Gates of the Arctic attained their full size at later ages than did bears from Alaska's southern coastal populations.[8] In addition to being smaller, Gates of the Arctic bears produced and successfully reared fewer cubs than did coastal populations.[9]

Small size, slow growth, and low reproductive output seen in Gates of the Arctic bears are all indic-

▶ While brown bears and humans both occur at low densities in the park, they do cross paths. (NPS/Claire Dal Nogare)

Based on bear movement data from GPS collars, streams were newly documented as anadromous fish habitat. Further research showed that interior brown bears that eat salmon have larger home ranges than those that do not, due to the large distances they cover between salmon streams and their denning areas high in the mountains. (NPS/Matthew Cameron)

ative of living in an environment with low productivity and limited availability of nutritious food items. While it was important that we document these traits of the Brooks Range bears, the results were not surprising. They are trademarks of the Arctic ecotype of interior brown bear populations, which differentiates them from their coastal brown bear relatives.[10] The Arctic is a low-productivity environment in general, and rough, interior mountainous regions here are even less productive than other continental areas where interior brown bears are found. We also confirmed that these bears are food limited or nutritionally deficient in an analysis of gene transcriptions (the process of copying information found in the animal's DNA).[11]

One of the ecological questions we investigated was, how do brown bears in Gates of the Arctic navigate this less productive system? We had assumed that these bears would adapt by having larger home ranges (the area they used during their active season to acquire resources scattered at low density across their landscape). We did, indeed, find that bears in the park, on average, had large home ranges, especially compared to their coastal relatives. However, it appeared that not all the bears were doing the same thing. Some bears, especially females with cubs, actually had fairly small home ranges (159 miles[2] [411 km[2]]), no bigger than those of moose living in the area.[12] These bears tended to stay up in the mountains all summer, likely foraging on new growth from grasses and forbs, roots, berries, ground squirrels, and the occasional moose or caribou. Others bears had among the largest home ranges ever reported for brown bears anywhere in the world, with a few male bears traversing an area larger than the State of Delaware (greater than 2.5 million acres [10,000 km[2]]) during the active season.[13] Interestingly, some the large home ranges were not bounded like a typical home range; rather, they appeared to be directional movements to certain areas. This likely highlights how far a bear must travel to find adequate food resources.

Many of these movements led to streams where bears ended up spending a substantial amount of time. We had received pilot reports of bears congregating and fishing along the Noatak River, in the western region of the park, but none along the southern flanks of the Brooks Range where these bears were going. Despite the fact that many of these streams did not appear on the official map of anadromous streams (waterways in which salmon return to spawn), we suspected that bears might be using the streams to fish. With this in mind, we used the GPS tracks of brown bears to identify streams where they might have been fishing for salmon, many of which were not previously documented as anadromous. During our first observational outings, in 2016 and 2017, we documented interior brown bears successfully fishing for chum salmon in all 3 of the major river systems flowing out of the park:

the Noatak, Kobuk, and Koyukuk Rivers.[14] With this confirmation, we realized that the existing map of anadromous streams needed to be updated. Our team reached out to the Alaska Freshwater Fish Inventory program, which completed new surveys in the region in 2018. This collaboration was extremely fruitful, adding hundreds of newly mapped miles (over 1,000 km) of streams to the state's anadromous waters catalog.[15] Isotope analyses of bear hair samples provided additional evidence that many interior brown bears in this region of the Brooks Range were consuming substantial quantities of salmon hundreds of miles from the coast. Over 75% of the meat they consumed in late fall was salmon.[16]

While having GPS-collared interior brown bears leading researchers to "new" salmon streams was an unexpected and important scientific discovery in and of itself, it also has very practical management implications.[17] Adding a stream to the anadromous waters catalog adds requirements along roadways for structures that adequately allow for fish passage. The Ambler Road is proposed to cross a number of these newly documented anadromous streams; if not for our research and collaborations, the road might have prevented fish passage and caused the loss of important fish habitat. That loss could have easily impacted bears that rely on fish to survive in this low-productivity environment.

The importance of salmon to bears and their environment is difficult to overstate. Salmon ma-ture and feed in the ocean and return to freshwater systems to spawn and die, thereby transferring marine nutrients to riverine systems. Bears (along with other terrestrial and avian species) feast on the temporarily available salmon; thus, marine nutrients are assimilated into the terrestrial environment. These nutrients are further distributed by bears as they urinate and defecate, providing important nutrients to growing plants in an environment that is nutrient-poor.[18] We found that salmon consumption was greater in large males but was not related to the size of females or the percentage of body fat of individual bears.[19] We believe this means 2 things: (1) larger-bodied males are likely better at finding and securing the prime salmon-fishing locations, and (2) brown bears in Gates of the Arctic can achieve similar body condition by being ecologically plastic. This means that individual bears can behave differently (that is, occupy different ecological niches) to achieve similar results in body condition.[20] In this case, the result is to get fat enough to reproduce and survive their long denning period. With individual bears occupying slightly different ecological niches, the niche used by the greater population is expanded. New techniques show that the breadth of niches used by brown bears may be larger than we previously thought.[21] Bears occupy different niches by eating different things in different amounts, at perhaps different times of the growing season, which can alter the size of their home ranges. For example, bears that focused their foraging efforts on salmon had, on average, 75% larger annual home ranges than bears that focused on vegetative resources.[22]

This finding of salmon-eating interior brown bears having larger home ranges was the exact opposite of what one might predict. Coastal brown bears, like those in Katmai National Park and Preserve and Kodiak National Wildlife Refuge, have some of the smallest home ranges among all brown bears. Salmon are so abundant and available for long periods of time in these places that bears do not have to travel far to get all the nutrients they need. In Gates of the Arctic, however, salmon resources are typically only available relatively far from bears' preferred denning and early summer habitat and for shorter periods of time. Thus, salmon-fishing bears need to travel further to this resource and then return back to their denning areas in the mountains, which leads to large annual home ranges. These movements outside the park can expose bears to risks, including sport hunting.

While GPS collars provide extensive information on the movements of an individual bear and can identify areas where bears concentrate their use, they do not shed any light on how many other bears might be doing the same thing. To address this, we also conducted a genetics study along 2 salmon

Bears in the north must travel long distances between salmon streams and their dens to take advantage of this important resource. (NPS/Matt Harrington)

streams where collared bears congregated, to identify how many other bears were fishing along these sections of stream during the chum spawning season. We extracted genetic samples from bear hairs that were collected from tiny snares we set along heavily used bear trails that ran along the banks of the streams. Based on our results, we estimated that 15 and 24 interior brown bears, respectively, fished along the 2 approximately 4-mile sections of salmon stream.[23] The estimate is impressively high for both streams, considering the low density at which interior brown bears occur in Gates of the Arctic (5.4–11.9 bears/100 square miles; 21–46 bears/1,000 km²),[24] and it highlights the strong dynamic between salmon and bears in this system.

Winter comes early in the Arctic, especially for mountain-dwelling interior brown bears. By late September, some salmon-fishing bears continue to feed on spawning chum at lower elevations, while other bears that live in the mountains year-round are the first to enter their dens. Food availability appears to be a key driver of when bears enter their dens. On average, bears remained in their dens for nearly 7 months, and we observed sows with new-

◄ In this extreme environment with a scarcity of food, bears remain in their dens for nearly 7 months of the year. Some sows with newborns stay in the den for 8 months—among the longest denning periods ever recorded for brown bears. (NPS/Matthew Cameron)

born cubs staying in their dens for almost 8 months. The observed denning durations are among the longest ever recorded for a brown bear population and highlight the importance of foraging effectively during the short growing season so bears can accumulate their annual nutritional needs for survival and reproduction in just 4–5 active months.

We found that Gates of the Arctic bears dig their dens on steep (>31°), well-drained, high-elevation, lee-facing (protected from the wind) slopes where snow depth is greatest throughout the winter. These features likely help keep bears dry and well insulated from the long, cold winter.[25] Bears rarely reuse the same den, likely, in part, because of the high rate of den collapse that occurs during the summer months. However, we learned that bears in this region often dig their den near their previous year's den site.[26]

While our primary research focus was on learning about the ecology of these rarely studied Arctic ecotype interior brown bears, we also wanted to collect baseline data on the population before development occurred in the region. We documented that bears in the park have a relatively low prevalence of bacterial, viral, and parasitic agents, as we expected for this very remote population. However, 14% of the bears sampled had been exposed to a canine distemper, a viral disease that affects a wide array of bodily systems in a wide range of mammals.[27] These bears were the only population studied that had

evidence of canine distemper, which potentially could be related to the historical and current use of sled dogs in the region. Levels of lead and cadmium in the blood of these bears were higher than those in other Alaska populations but much lower than those found in Scandinavian bears.[28] Low mercury levels may be related to bears' lower consumption of salmon relative to the other Alaska populations sampled. Finally, species diversity within individuals' gut microbiome was lower than that found in Katmai and Lake Clark, which is likely related to the relative lack of forage diversity in the Arctic compared to those more coastal parks.[29]

In summary, interior brown bears in Gates of the Arctic National Park and Preserve are living near the edge of their range in a wilderness area characterized by extreme environmental conditions and low habitat productivity. This contributes to their small size, low density, and low productivity. Individual bears deal with these limitations in different ways: some bears stay up in the mountains year-round foraging on grasses, roots, and berries, while

Brown bears are well adapted for digging. Bears normally dig their dens on steep slopes at high elevations, protected from the wind, and near their previous year's den site. (NPS/Kyle Joly)

Brown bears in the Arctic have responded to living in a harsh environment by having very long denning times, very large home ranges (or very small ones, depending on the location), small body size, low productivity, low densities, and by finding good fishing locations. (NPS/Erika Jostad)

other bears make forays out to salmon streams later in the summer, only to return to the mountains to den. Some bears have extremely large home ranges and others have relatively small ones. All this variability expands brown bears' overall ecological niche in the remote Arctic ecosystem and may promote resiliency over the long term. One thing they do have in common is that they all have very lengthy stays in their dens. Bears have developed a range of strategies to cope with their harsh environment, but the cumulative impacts climate change and expanding human development will have on the population remain uncertain.

FUTURE NEEDS

Gates of the Arctic National Park and Preserve has the longest, most complex enabling legislation of any park unit created by ANILCA, stemming from a legislative compromise to allow for surface transportation across the Kobuk River section in the southwestern corner of the preserve. This would allow road access from the existing road system to the east to the Ambler Mining District. A permit for the right-of-way was granted in January 2021. It was revoked in 2024, but debate continues. Therefore, the most immediate need for brown bear management and research in the park is to complete the analyses of the 2014–2017 GPS data. The most critical piece is to examine habitat use along the Ambler Road corridor to identify where *not* to plan

infrastructure. That is, are there places where the corridor can be rerouted to lessen the impact on bears and the salmon resources they rely on? This information could also be used to avoid placing construction camps, road maintenance facilities, ranger stations, and campgrounds in areas with high use by bears to minimize human-bear interactions along the corridor. Access, like boat launches, hiking trails, and fishing locations, could also be informed by such analyses. Access, which will be much greater once the Ambler Road is developed, will likely also affect bear harvest.

In general, there is no detailed understanding of the total human-caused removals (or mortalities) of interior brown bears in Gates of the Arctic, which include legal harvest, defense of life and property (DLPs) take, and poaching. Increased access to this currently incredibly remote landscape will most likely lead to additional removals by people. Because large swaths of high-quality bear denning habitat are located in the park, bears that travel outside park boundaries to fish for salmon could interact with the Ambler Road and be more at risk. In other words, impacts from outside the park may affect the bear population within the park. Given the low productivity of the system and the bears living here, small increases in mortalities could have substantial impacts on the overall population.

To truly understand what impact increased mortality might have, a solid understanding of

Faced with the uncertainty driven by climate change and development, perhaps the best way to conserve brown bears in the Arctic is by building appreciation of them. (NPS/Erika Jostad)

how many bears are living on the landscape is needed. Continuing the population surveys that were undertaken in 2010 and 2018 will help us better understand the population trajectory of brown bears that reside in Gates of the Arctic. This work covers only the northeastern portion of the park. Trying to understand the population trends on the southern flanks of the park is an important management consideration. This area has more shrubs and forested area, which makes estimating brown bear abundance much more difficult (which is one of the reasons why the tundra-dominated northeastern section was chosen for population monitoring). Estimating brown bear abundance in this environment may require hair snaring, camera trapping, DNA sampling, or novel methods—and a lot of time, money, and effort.

Another important future management decision that directly follows from our research is how to deal with the numerous waterways we discovered that have high levels of brown bear activity. In other words, how does one protect vital salmon-bear streams while not highlighting them? Highlighting the places where bears are can lead to increased visitation, disturbance to bears (and thus loss of foraging time), human-bear incidents, DLPs, hunting, and poaching. Within Gates of the Arctic, seasonal closures could be implemented, as has been done in other conservation areas (e.g., Denali National Park and Preserve; Yellowstone, Glacier, and Grand Teton National Parks). Outside the boundary of Gates of the Arctic, decisions are more complicated and management lies with other agencies, though the impact on bears denning inside the park could still be high. Sharing the location of these critical bear habitats without acquiring appropriate mitigation or conservation measures could be counterproductive for the park's brown bear population.

Of course, one of the most powerful uses of the 2014–2017 brown bear data is as a pre-road baseline. These data are vital to assess potential impacts of the construction and operation of the Ambler Road.

Securing funding for, implementing, and analyzing effects on bears after the road construction is completed will be critical for the adaptive management of the road itself, associated human access and use, and the brown bears in this region.

Along with construction of the Ambler Road, we believe climate change is likely to have large impacts on brown bears. Warmer summers may increase vegetative productivity, providing bears with more food resources. The longer growing season will likely extend the active season (the amount of time they are not in their dens), which would allow them to consume more calories, grow larger, and successfully rear more and possibly larger offspring. Warming ocean waters in the Arctic may allow more salmon to reach them. However, if river waters get too warm during the summer, they may run low on oxygen and salmon may do poorly. Already, poorly oxygenated waters have been suspected in large fish die-offs in the Kobuk River, so depleted salmon runs are a real concern for bears.[30] Another factor in which a little change may benefit Gates of the Arctic bears but a large change might not is annual snowfall. Less snow may help extend the growing season of vegetation in the Arctic and extend the active season of bears; however, Brooks Range bears rely on snow to help insulate their dens during the long winter.[31] So, a lack of snow could be detrimental during this critical period. Thus, a much better understanding

▸ A bear naps while guarding what's left of a caribou carcass. (NPS/Matthew Cameron)

of how climate change will impact interior brown bears is needed.

Faced with a future of uncertainty from both climate change and anthropogenic development, we recommend a cautious approach to this low-density, low-productivity population. In the end, perhaps the greatest need of bears in Gates of the Arctic is to have people appreciate and understand their ecology so they want to conserve the bears. That is the most important key to conserving brown bears in perpetuity. We hope our research and this chapter help to do just that.

NOTES

1. Nelson, Mautner, and Bane, *Tracks in the Wildland*; Loon and Georgette, *Contemporary Brown Bear Use in Northwest Alaska*; Anderson et al., *Kuuvaŋmiut Subsistence*.
2. Nelson, Mautner, and Bane, *Tracks in the Wildland*; Loon and Georgette, *Contemporary Brown Bear Use in Northwest Alaska*; Anderson et al., *Kuuvaŋmiut Subsistence*.
3. Nelson, Mautner, and Bane, *Tracks in the Wildland*.
4. Glover, *A Wilderness Original*.
5. Glover, *A Wilderness Original*.
6. DeVoto, *The Journals of Lewis and Clark*.
7. National Park Service, *Management Policies 2006*.
8. Hilderbrand et al., "Body Size and Lean Mass of Brown Bears across and within Four Diverse Ecosystems."
9. Hilderbrand et al., "Influence of Maternal Body Size, Condition, and Age on Recruitment of Four Alaska Brown Bear Populations"; Hilderbrand et al., "Brown Bear (*Ursus arctos*) Body Size, Condition, and Productivity in the Arctic."
10. Cameron et al., "Body Size Plasticity in North American Black and Brown Bears."
11. Bowen et al., "Using Gene Transcription to Assess Ecological and Anthropological Stressors in Brown Bears."
12. Joly et al., "Factors Influencing Arctic Brown Bear Annual Home Range Sizes."
13. Joly et al., "Factors Influencing Arctic Brown Bear Annual Home Range Sizes."
14. Sorum, Joly, and Cameron, "Use of Salmon (*Oncorhynchus* spp.) by Brown Bears (*Ursus arctos*)."
15. Cathcart and Giefer, "Fish Inventories of the Upper Kobuk and Koyukuk River Basins."
16. Mangipane et al., "Dietary Plasticity and the Importance of Salmon to Brown Bear (*Ursus arctos*) Body Size and Condition."
17. Sorum et al., "Salmon Sleuths."
18. Levi et al., "Community Ecology and Conservation of Bear-Salmon Ecosystems."
19. Mangipane et al., "Dietary Plasticity and the Importance of Salmon to Brown Bear (*Ursus arctos*) Body Size and Condition."
20. Hilderbrand et al., "Plasticity in Physiological Condition of Female Brown Bears across Diverse Ecosystems."
21. Rogers et al., "Splitting Hairs."
22. Joly et al., "Factors Influencing Arctic Brown Bear Annual Home Range Sizes."
23. Sorum et al., "Pronounced Brown Bear Aggregation along Anadromous Streams in Interior Alaska."
24. Schmidt et al., "Brown Bear Density and Estimated Harvest Rates in Northwestern Alaska."
25. Sorum et al., "Den-Site Characteristics and Selection by Brown Bears."
26. Sorum et al., "Den-Site Characteristics and Selection by Brown Bears."
27. Ramey et al., "Exposure of Alaska Brown Bears (*Ursus arctos*) to Bacterial, Viral, and Parasitic Agents."
28. Haugen, "Blood Concentrations of Lead (Pb), Mercury (Hg), and Cadmium (Cd) in Scandinavian and Alaskan Brown Bears."
29. Trujillo et al., "Intrinsic and Extrinsic Factors [*sic*] Influence on an Omnivore's Gut Microbiome."
30. von Biela et al., "Evidence of Prevalent Heat Stress in Yukon River Chinook Salmon."
31. Sorum et al., "Den-Site Characteristics and Selection by Brown Bears."

10

KATMAI

THE GREATEST CONCENTRATION OF BEARS ON THE PLANET

Kelsey Griffin, Michael Saxton, Joy Erlenbach, Leslie Skora, Troy Hamon, Laura Stelson, and Grant V. Hilderbrand

HISTORY

Ancestors of the Alutiit-Sugpiat and Aleut peoples have lived along the Katmai coast for at least 9,000 years. The abundance of rich, seasonal food sources along the coast supported some of the largest villages seen in Alaska before the arrival of Russian settlers in the 1780s.[1] While some settlements maintained substantial autonomy in the face of growing Euro-American influences in the region, others were coerced into servitude by a growing number of artels (corporate villages) or traders, who sought Indigenous hunting skills and knowledge of the region to pursue fur animals such as fox and sea otter. Pelts were increasingly traded for credit toward manufactured food and goods.[2] As men spent more and more time away from the villages, women maintained their subsistence traditions at home. When fur animals grew scarce from over-hunting, many turned to seasonal employment opportunities in the growing industries of fish salting and canning, leaving many households and communities permanently transformed.[3]

A bear pounces on a salmon in the Brooks River in Katmai National Park and Preserve. (NPS/R. Jensen)

https://doi.org/10.5876/9781646427116.c010

The abundance of bears in Katmai National Park and Preserve did not receive attention until 1989, when scientists and others began investigating the effects of the Exxon *Valdez* oil spill. (NPS)

The Brooks River area has been used regularly by Indigenous communities for nearly 5,000 years, with use prior to that point largely transitory hunting of migrating caribou. When the Brooks Falls began to form about 5,000 years ago, it created excellent conditions for capturing migrating salmon, which led to year-round habitation and the formation of local communities.

The Novarupta volcanic eruption in 1912 prompted settlements of people living on the coast to evacuate as their communities were buried in feet of ash. The establishment of Katmai National Monument in 1918, just 6 years after the eruption, included coastline from Kashvik Bay north to Kuliak Bay. The establishment of the national monument prevented those who had grown up within its

KAMISHAK BAY

ALAGNAK WILD RIVER

Alagnak R.

Kukaklek Lake

Moraine Cr.

Funnel Cr.

Battle

Nonvianuk Lake

Kulik L.

American Cr.

Strike Cr.

Kamishak R.

King Salmon

KATMAI NATIONAL PARK AND PRESERVE

Lake Colville

Lake Grosvenor

Mount Douglas

Fourpeaked Mountain

Douglas R.

Naknek Lake

North Arm

Handscrabble Cr.

Savonoski R.

Swikshak Bay

Brooks Camp

Iliuk Arm

Lake Brooks

Ukak R.

Rainbow R.

Serpent Tongue Gl.

Mount Steller

Mount Denison

Hallo Bay

Hallo Gl.

Swikshak Bay

Salmon R.

Contact Cr.

Angle Cr.

Valley of Ten Thousand Smokes

Mount Mageik

Katmai R.

Kukak Bay

Becharof Lake

Katmai Bay

Dakavak Bay

Kashvik Bay

SHELIKOF STRAIT

SHELIKOF

Miles
0   20

N

Katmai National Monument was established in 1918, 6 years after the massive Novarupta volcanic eruption. In 1931, the boundary was expanded to include Brooks Camp and more than half of the current coastline included in the park today. In 1980, it was expanded again and became Katmai National Park and Preserve as part of ANILCA.

boundaries from returning and reforming permanent settlements. But the Brooks River, as well as other regions within the park, continued to be used for subsistence by descendants of those who lived in the region prior to the Novarupta eruption.

In 1931, the monument boundary was expanded to include the Brooks Camp area along with more than half of the coastline that is part of the park today. The expansion also added the protection of brown bear populations to the enabling legislation. The boundary was further extended in 1942 to include all islands within 5 miles (8 km) of the coast. Despite much of the coastline being included as parkland since 1931, visitation to the coast was limited. Activity came primarily from the clamming and salmon-fishing industry, with canneries established at Kukak Bay and Swikshak Bay beginning in the mid-1920s until the 1950s.[4] Although commercial clamming has been discontinued, there is some commercial fishing of pink salmon along the coast.

From its creation until 1971, Katmai was administered by Mount McKinley National Park (now Denali). Despite the expansion to incorporate the area of Brooks Camp in 1931, the administration of the monument from Mount McKinley and a lack of funding meant that no rangers visited the area until 1937. The establishment of the Bureau of Fisheries laboratory and weir in 1940 brought a consistent government presence to the area during the summer that improved the protection provided

to bears and created an artificial feeding pool below the weir, which some bears were reported to visit nocturnally.[5]

Even though it was included in the national monument with the 1931 expansion of the boundaries, it wasn't until 1950 when the National Park Service (NPS) reached an agreement with Ray Peterson to establish a fishing lodge on the Brooks River that any permanent NPS personnel were stationed within the national monument. During this early establishment of Brooks Camp, bears were uncommonly seen in the area because they were considered competitors for food, were hunted, and were therefore leery of humans in the area. Bears were so rare that lodge managers made note in their logbooks when tracks were seen on the beach. At this time, the lodge was used exclusively as a fishing lodge, and photos show lodge patrons fishing right at the Brooks Falls, where we might see upward of 30 brown bears competing for space today. In the following decades, due to a revised management philosophy that emphasized the intrinsic value of bears and looked to minimize human-caused bear deaths, bears became more common. In time, NPS employees struggled with the same issues of food and garbage storage that were familiar to some parks in the Lower 48 states.[6]

In 1971, in response to growing problems with food-conditioned bears, the park began providing all visitors with bear-safety talks upon their arrival at Brooks Camp, a practice that continues today in a modified form. In addition, new measures were taken to try to limit bears' access to human food, as food-conditioned bears were regularly accessing trash, even when the park and lodge boated it across the lake and buried it. It reached the point where employees transporting trash across the lake would be met by bears waiting for the trash to be offloaded, causing the employees to fear for their safety.[7] Further measures introduced throughout the 1970s and 1980s included incinerating trash, limiting locations where food could be prepared and consumed, and requiring proper storage of food. These efforts were effective, and the frequency of bears obtaining food dropped.

The Alaska National Interest Lands Conservation Act (ANILCA), in 1980, expanded Katmai, with new park and preserve areas. Like most national parks in the Lower 48 states, hunting is not permitted in the original park. Subsistence hunting is allowed in the new park additions as well as the preserve areas. Recreational ("sport") hunting is also allowed in the preserve areas.

It wasn't until 1989, in the aftermath of the Exxon *Valdez* oil spill, that broader attention was brought to the Katmai coast. As scientists and others worked to understand the impact of the oil spill, they discovered an abundance of bears and research opportunities. Eventually, guides began bringing people to "bear view" on the coast, arriving primarily from

Anchorage, Homer, Kodiak, and King Salmon. The closest communities to the Katmai coast are King Salmon, located on the other side of the Aleutian Mountain Range, and Kodiak Island to the east across the Shelikof Strait. Access to the Katmai coast is limited to boat or plane.

As bear viewing gained popularity, more people also began visiting Brooks Camp with a primary focus of that activity, a transition from the previous decades of focus on fly-in fishing trips. The grow-

Most people arrive at the park by plane. Nearly 7,000 people visit the Katmai coast every summer, with as many as 400 people arriving at Brooks Camp per day. (NPS/ Kyle Joly)

ing popularity of bear viewing closely followed an increase in the number of bears using Brooks River. As the area became more crowded, the park constructed elevated viewing platforms for visitors to watch bears from a safer space, keeping people and bears physically separated at areas of high bear use. Over time, the extent of these platforms increased, from a single elevated platform at the falls to the platform to boardwalk complexes that exist at both the falls and the mouth of the river, where bears congregate in the fall to feed on dying salmon after the spawn.

Today, almost 7,000 people visit the Katmai coast each summer, with just a few sites receiving most of the visitation. During peak season, as many as 400 people may arrive per day to visit Brooks Camp, with nearly 16,000 visitors across the season. Most arrive by float plane, with over 30 float planes landing and taking off regularly throughout the month of July to deliver visitors. Boats and small airplanes can be seen dotting sandy beaches and spots of calm water, bringing visitors for wildlife and glacier viewing, fishing, and camping. The most popular activity is bear viewing (more than 75% of visitation), with Hallo Bay and Geographic Harbor 2 of the most widely known coastal bear-viewing destinations.[8]

ENVIRONMENT

Bears in Katmai are larger than their inland relatives and occur at higher densities. The abundance of food resources available to bears in Katmai may be part of what makes them differ in size, behavior, and density compared to their inland relatives. Tolerance of being in close proximity to other bears may also extend to people. That tolerance, a lack of hunting, and habituation offer some of the best bear-viewing opportunities in the world.

Bears share the Katmai coast with numerous wildlife, including seabirds, shorebirds, and mammals such as wolves, moose, foxes, and ground squirrels. Seals, otters, and sea lions haul out on nearshore islands and tidal sandbars. Four species of salmon spawn in coastal streams: chum, pink, coho, and sockeye.[9] Barnacles, mussels, clams, crabs, and other inter-tidal organisms are abundant; and whales and other marine life occasionally wash up on Katmai shores. There is little human presence and the landscape is largely wilderness; there are no roads, villages, or permanent full-time residents.

Brooks Camp, situated at the mouth of the Brooks River, is a 1.5-mile-long (2.4 km) waterway connecting Naknek Lake to Brooks Lake. Much of the river is surrounded by forest, with dominant species white spruce, alder, willow, and birch, with stands of cottonwoods abundant through the developed areas of Brooks Camp and the campground. The

Bears in Katmai are larger than inland brown bears and live in closer proximity to one another. (NPS/Lian Law)

Brooks River varies in width, from approximately 80 feet to 165 feet (24 m to 50 m), though it widens at the mouth. Where it enters Naknek Lake, seasonal wetlands surround the river, and the forest retreats away from the banks. Brooks Falls is a 6-foot-high (2 m) waterfall, approximately halfway up the length of the river. It creates a partial barrier to salmon migration that results in large numbers of fish pooling at the base of the falls and creates a concentrated feeding opportunity for bears.

A part of the Naknek River drainage, the Brooks River is an important spawning and nursing ground for the Bristol Bay salmon population, the largest sockeye salmon fishery in the world. To reach the Brooks River, salmon migrate approximately 60 miles (96 km) up the Naknek River and through Naknek Lake. Some salmon spawn in the Brooks River, while many others move through the river and continue migrating to spawning locations along the banks of Brooks Lake and small tributary streams. As spawning salmon return from the ocean each year, it is one of the first easy places for bears to acquire fish. All 5 species of Pacific salmon are found in the Brooks River, but sockeye salmon are by far the most numerous. The roe and flesh of the salmon after the spawn provide a bonanza for other fish species such as rainbow trout, lake trout, grayling, and Dolly Varden. This abundant food source is the reason the Brooks River is highly regarded as a world-class fishery.

The Katmai coast spans approximately 500 miles (800 km) on the Pacific side of the Alaska Peninsula. It is composed of vast sand and cobbled beaches, salt marshes, and steep rocky cliffs. Dynamic and exposed shorelines are dotted with occasional islands and sheltered coves and harbors. The coast retains a wild remoteness, evident not only through its wilderness designation but also its unique habitats. Lowland marshes, grass-forb meadows, streambeds, occasional stands of birch, cottonwood, and Sitka spruce, as well as tundra-like vegetation give way to mid-elevation shrub and alder thickets. Mountains rise sharply with glaciers, crater lakes, and dormant volcanoes of the Aleutian Range, with the highest, Mount Denison, reaching approximately 7,600 feet (2,316 m).

Vegetation, including sedges, goose tongue, angelica, cow parsnip, and lupine, as well as berries such as salmonberry, nagoonberry, crowberry, blueberry, and cranberry, are found along the coast. Many of these resources, including both marine and terrestrial sources, are important foods for bears. The bears of Katmai can be seen by the dozen grazing on salt marshes in the spring and early summer and concentrated on salmon streams through the summer and fall. Brooks Camp is at an elevation of 50 feet (15 m) but is immediately adjacent to Dumpling Mountain, which climbs to 2,490 feet (759 m). This and several other nearby mountains provide denning habitat for brown bears.[10] High- and low-bush cranberry, as well as nagoonberry, can be found in the lower elevations near the river and lakes, while blueberries and crowberries are found in the alpine tundra zone on the mountains.

The temperate marine climate of the coast contributes to milder temperatures compared to other parts of Alaska. Wet weather is common, and storm systems that move up and down the Aleutian chain bring wind and rain in the summer and snow in the winter. Strong winter storms erode cliffs and shift beaches. While climate data specific to Brooks Camp are not available, the community of King Salmon, approximately 30 miles (48 km) west, drops well below zero in the winter, with average temperatures around 25°F (4°C). Summers are cool, often rainy and windy, with average temperatures around 65°F (18°C).[11]

MANAGEMENT CHALLENGES

The primary management challenge at Katmai is undoubtedly bear viewing by the many visitors who arrive each year. Viewing a bear is a top priority for visitors from outside Alaska, and Katmai is one of the premier locations to have this experience. Climate change is affecting all aspects of Alaska, including the ecology of brown bears. How bears will handle these changes is a management

The salmon runs are the main attraction at Katmai National Park and Preserve. They allow bears to gain fat stores and tolerate other bears competing for food. (NPS/Lian Law)

concern for the park because of the implications for bear populations and bear-viewing opportunities. Finally, bear harvest for sport and subsistence purposes allowed in the preserve presents a management challenge as we balance consumptive and non-consumptive uses.

Bear Viewing

Since 1998, human use of one coastal area (Hallo Bay) has increased 1,000%, from 300 to >3,000 people/year (1998–2015).[12] In other locations along the coast, cruise ship activity has increased, with small cruise ships visiting some of the deeper bays. Though these visitors can be numerous, documentation of their visitation is lacking because these passengers often remain in marine waters just adjacent to parklands but may never technically enter the park. Photographers, filmmakers, campers, and researchers also use the coast and often stay for multiple days or weeks.

Visitor safety is also a primary management concern. Visitors concentrate in areas with high bear activity, resulting in groups of people dispersed in an area where bears are feeding. Depending on the site, a guide may take the group along trails and across streams to watch bears in sedge meadows and along riverbanks, beaches, and inter-tidal areas as bears fish or dig for clams. Direct effects such as trampling of vegetation and behavioral effects (e.g.,

increased vigilance) may result from human use of bear habitat. Documented behavioral impacts of human presence on wildlife can be numerous, including spatial avoidance, temporal avoidance, and changes in the amount of time spent in a habitat, in the number of animals present, in sex/age class of animals in a habitat, and in activity budgets, including increased vigilance or energy spent on fleeing.[13] Research also shows that predictability of human presence can minimize some of these behavioral effects on bears; however, since bear viewing is currently structured on the coast, visitation is only semi-predictable. Only 1 site (Geographic Harbor) manages the space used by visitors, and the remainder of the coast is free of spatial or temporal restrictions for visitation. Predictable visitation is also a challenge, as weather and tides dictate much of how visitors can use the coast.

Likewise, the biggest management challenge facing Brooks Camp is the rapid increase in visitor use. The rate of growth in visitation at Brooks Camp exceeds that seen across the Alaska region and nationally. Since 2010, each year has seen a record number of visitors, except for 2020, when the Covid-19 pandemic resulted in a dramatic drop in the number of visitors. When the camp was

▶ Visitor safety is a primary concern. On their arrival at Brooks Camp, visitors are briefed to know what to do if they meet a bear on a trail. (NPS/Lian Law)

established, bear populations were much lower; therefore, no thought was given to the challenges of coexisting alongside these animals. Today, the bear population on the Brooks River is at a historic high. With human visitation reaching a new record again in 2021, the increase in both populations creates challenges for those tasked with managing people and bears in this area.

The location of Brooks Camp, in a prime travel corridor on a spit between the river and Naknek Lake, has long been recognized as a source of conflict between people and bears. This led to the determination as part of the 1996 Development Concept Plan (DCP) for Brooks Camp to move camp to the south side of the river and further into the woods. In addition to moving camp, the DCP established capacity limits on visitation to improve visitor safety, minimize impacts to wildlife, and maintain the quality of the visitor experience, with a cap of 205 visitors in July and 170 in September. Unfortunately, outside pressure has made progress toward these goals incremental, where possible at all. Since the plan was passed in 1996 and visitation has grown, the number of people in the area, particularly in July, greatly exceeds those previously determined capacity limits. A renewed push to either implement the existing DCP or develop a new management plan for the Brooks area is necessary to guide management.[14]

Unlike other areas of the park, most visitors to Brooks Camp are unguided as they hike to the falls or up to Brooks Lake. This means that people are interacting with brown bears without the benefit of an experienced guide to direct them in how to respond if a bear approaches. The lack of guided visitation results in significantly less predictable human behavior, determined by preexisting biases, fears, knowledge, and beliefs. To help address this issue, the park provides an orientation for all visitors arriving at Brooks Camp, briefing them on the basic procedures and how to respond in a bear encounter. While this has proven very successful, that success is largely thanks to the tolerance of the bears.

Despite these challenges, Brooks Camp has successfully kept bears and people safe from each other. In 70 years of operation, there have been remarkably few injuries from bears, and 1983 was the last year a bear was destroyed for management purposes in Brooks Camp. The most significant injury to a person occurred in 1966, when a camper prepared a fish and did not properly clean up. This was prior to the introduction of the electric fence at the campground, which was installed in 2000. After the camper had gone to bed, a bear came through the campground and bit him on the leg. The bear was chased off, and the camper required a stay at the hospital in Anchorage to recover from his injuries. Aside from that incident, a ranger sustained a minor puncture wound that was suspected to be from a bite after an incident with a bear on the old falls trail, and a visitor was knocked to the ground when 2 young bears chasing each other ran over him. The changes to food and gear management, significant improvements in visitor education, and introduction of the waste incinerator—all in the 1970s—have led to remarkably few negative encounters between bears and people at Brooks Camp.

There have been few reported serious human-bear conflicts on the coast and only 2 human deaths since records have been kept. A bear killed 2 people who were not part of a commercially guided tour in 2003, resulting in park staff responding to the incident and killing 2 bears. Only 1 other incident resulted in the death of a bear after it charged at a park staff member. Continued attention to strategies for managing increasing visitation and human-bear interactions will be important for the health and safety of brown bears and visitors to the Katmai coast.

The park has made efforts to address some human-use concerns through changes to park regulations: a minimum distance above the ground for airplanes, seasonal closures of overnight camping within salt-marsh habitat at Hallo Bay, designation of a viewing area at Geographic Harbor, limits to group size, and the number of days a party can spend in 1 campsite. In addition, people may not approach a bear or occupy a position within 50 yards (50 m) of a bear using a concentrated food source such as an animal carcass or spawning salmon anywhere in Katmai.[15] In high bear density areas,

encounters that are much closer than 50 yards (50 m) do occur, sometimes inadvertently.

To minimize the chances of bears acquiring human food, all unattended food and garbage must be secured in approved bear-resistant food containers, and the use of electric bear fences is encouraged. Some sites require the removal of all human waste and toilet paper. In addition, people are instructed to report all incidences of bears in contact with camp, gear, or food or those behaving in an aggressive manner. The use of a firearm, bear spray, flare, or other deterrent is also to be reported to the park. In addition to park regulations, commercial operators are encouraged to follow the best practices for viewing bears developed by the National Park Service with the State of Alaska and other stakeholders.[16]

A small staff and a vast park area to cover make enforcement of regulations and education difficult on the coast. Visitors to the coast may never encounter park staff members and sometimes do not even realize they are in a national park. Interpretation, communication of park information, and enforcement of rules and regulations come in most cases from tour operators who guide visitors into the park rather than from park staff. Many commercial operators have guides who have been operating in the park for years or even decades, who understand bear behavior, and who have a shared desire to protect park resources. However, because they are not park staff and there is no mandatory standardized training or enforcement, information presented to visitors can vary in content and quality among providers.

Climate Change

At the interface between the land and the sea, management challenges in coastal Katmai are multidirectional. Climate change and anthropogenic impacts including the redistribution and restructuring of vegetation communities, altered growing seasons, flooding, and erosion can have substantial terrestrial effects. Warmer ocean temperatures and ocean acidification can have substantial marine impacts, as they did in the Gulf of Alaska during the Pacific heat wave from 2014 through 2016, when altered plankton and forage fish availability resulted in seabird and marine mammal die-offs.[17] Continued warm-water anomalies have the potential to increase toxic algal blooms and the frequency of food-chain disruptions, with broad impacts on fish, invertebrates, seabirds, and the species that rely on them, including bears. Pollution, marine debris, micro-plastics, increased vessel traffic, commercial development, and oil spills are also management concerns, as they all have the potential to disrupt both marine and terrestrial food webs.

A potentially catastrophic challenge posed by the changing climate is how it might impact salmon populations. Salmonids are temperature-sensitive in both their spawning environments and warmer annual temperatures; in addition, reduced glacial melt feeding spawning streams may affect the health of this important population. As the keystone species in the Bristol Bay ecosystems, salmon directly and indirectly support hundreds of species and act as a conveyor belt, transporting nutrients inland. The abundant prey provided by migrating and spawning salmon allows for brown bear densities found in few other places on Earth. What impact rising temperatures in the spawning and nursery environments will have is as yet unknown, as are the unknown effects of rising ocean temperatures and the associated acidification of the marine habitat. One indication of what may be to come was seen in 2019, when a record heat wave hit the Alaska Peninsula in early July. As temperatures rose, the Brooks River warmed significantly above 68°F (20°C), the lethal temperature for sockeye salmon. Salmon that had just begun to migrate through the river moved back into deeper water in Naknek Lake and did not reenter the river for a week. This occurred shortly after large congregations of bears had arrived at the river to feed on salmon, and their absence led to tense conditions and several negative encounters between bears and visitors.

▶ A sow bear with her three cubs of the year keep a watchful eye on other bears. (NPS/Naomi Boak)

Harvest Management

Two kinds of harvest occur in Katmai's preserve portion. Sport hunting is managed by the state, in accordance with NPS regulations, and the park requires hunting guides that meet commercial-use guidelines. Subsistence also occurs in the preserve and is managed by the Federal Subsistence Board and the NPS. To prevent over-harvest, the NPS monitors brown bears in Katmai Preserve. This monitoring is challenging due to the park's remote location, harsh weather conditions, and difficult and expensive logistics. Bears can be hard to detect on the landscape; as a result, biologists have taken advantage of periods and places when bears congregate on the landscape.

The park has conducted on-the-ground bear and human monitoring during August salmon runs. Increased bear viewing has created concerns, as bears become habituated to human presence during the salmon spawning season while bear-viewing groups seek close encounters with bears. Then, during bear-hunting season, the hunting of habituated bears can become an ethical dilemma.

RESEARCH SUMMARY

Denning and Movement Patterns

In the early 1970s, biologist Will Troyer came to the Katmai region to study bear denning patterns and movement, kicking off the modern era of scientific research on the brown bears in the Brooks Camp area.[18] After completing his initial den surveys, Troyer conducted research on the population dynamics and demographics of the bear population on the Bristol Bay side of the Aleutian Range, helping establish survey protocols followed today by several management agencies.[19]

Troyer's work identified the status and location of various feeding congregation and denning sites throughout the monument and used radio collars to examine movement and dispersal of the bears that congregate at the Brooks River.[20] While this work was foundational for this population, it was not without significant limitations. Due to the size of the bears' necks in comparison to their heads, as well as the relative novelty of the technology (radio collars had only begun to be used to track wildlife in the previous decade), many bears quickly lost their collars, some within feet of where they were collared. In addition, infrequent survey flights limited the number of locations identified for each bear, and relatively few bears were included in the study. Despite these limitations, this work remains the most complete knowledge we have of movement patterns of bears on the Bristol Bay side of Katmai National Park and Preserve.

Human-Bear Interaction

In the 1980s, the park began a collaboration with university researchers with the intention of better understanding bear-human interactions at Brooks Camp. Dr. Barrie Gilbert, a professor of wildlife ecology at Utah State University, began a series of investigations at bear-feeding congregations and burgeoning bear-viewing locations throughout the State of Alaska, including Brooks Camp. Dr. Gilbert's initial investigations in Brooks Camp, conducted in collaboration with graduate student Anne Braaten, used observational monitoring to measure the impact visitors in the river corridor were having on the bears using the river as a concentrated food source.[21] After Braaten concluded her work, Dr. Gilbert brought in a new student, Tammy Olson, who continued the work started with Braaten. The work conducted by Gilbert, Braaten, and Olson found that increasing human use of the area was causing some bears to spatially or temporally avoid areas in response, altering bears' use patterns of the river.[22] Olson and Gilbert made note of the difference between what they term "habituated" or "non-habituated" bears in the impact people have on

Cubs cling to their mother as she crosses the river. (NPS/Lian Law)

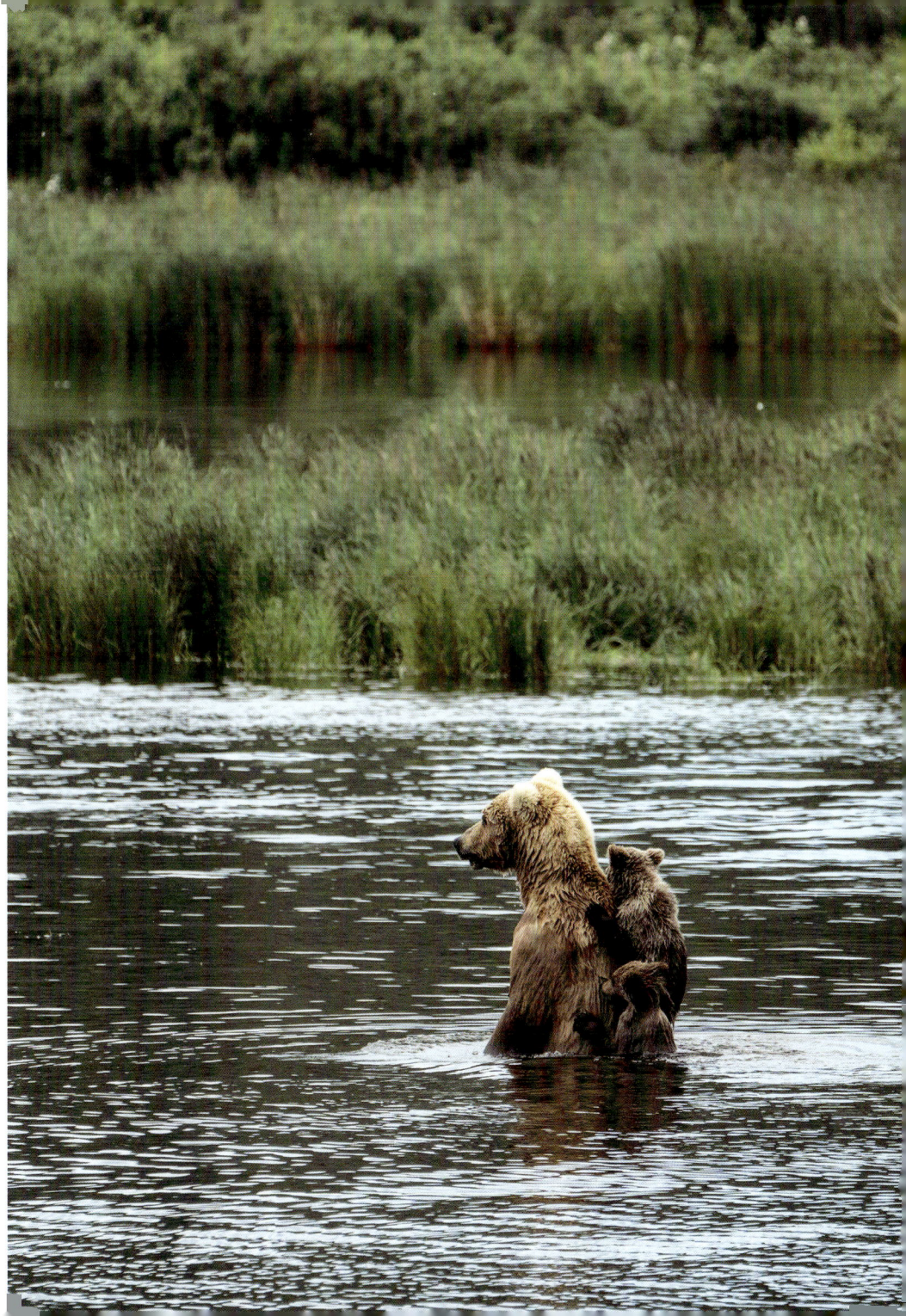

bears. They noted that while habituated bears show no change in behavior, there was a class of bears that were unwilling to tolerate the human disturbance.[23] These bears were modifying their behavior to avoid people who were using the river, either feeding nocturnally or leaving the area entirely. Olson and others also compared the impacts at Brooks River to Margot Creek, which has a similar feeding congregation but receives much lower visitation.[24] They found divergent use patterns of these 2 salmon streams, with the bears at Margot Creek displaying a uniform distribution of use throughout the day and the bears at Brooks far more likely to be active in the early morning and late evening, when human use of the river was lower.[25] After completing her graduate studies, Olson continued her work with the Brooks bears as the park wildlife biologist. In the 1990s, Brooks Lodge requested permission to extend its fall season operations by a week, from a historical closing date of September 10 to a new closing date of September 17. Olson was concerned that bears that entered the river after human use diminished would be excluded from the area if this change occurred. Despite those concerns, the later closing date was approved, and Olson used the opportunity to evaluate the impacts to bears during that brief window. She found that, indeed, many bears that were classified as non-habituated significantly decreased their use of the river in the fall after lodge operations were extended.[26]

While Olson and Gilbert were working on the Brooks River, Tom Smith, with the United States Geological Survey, was conducting similar work on the Kulik River, approximately 38 miles (61 km) away, also within Katmai National Park and Preserve. Smith found similar impacts, with bears spatially and temporally avoiding areas of high human use.[27] He also found that approximately 25% of bears that were involved in an interaction with people on the Kulik River left the river for several hours after the encounter, time in which they might otherwise have been building fat reserves for hibernation and reproduction.[28] In addition, bears that were using the river when people entered near them were far more likely to leave the river after a short time than were bears that did not encounter people in the river.[29] Smith also noted that bears that did use the areas of the river most frequented by people spent less time resting and more time moving.[30]

While the work of Olson, Gilbert, and Smith found detrimental impacts on bears of high levels of human use, other reports have documented mixed effects, with some age and sex classes benefiting from a level of tourism while other classes were less likely to use the area.[31] The differences may be a result of the level of human use, long-term effects of habituation, or the ways bears use the area. Despite the early reports of detrimental impact, ever-increasing human presence on the Brooks River has not yet resulted in a reduced number of bears; in fact, as human use has increased since the mid-1980s, so has the number of bears documented using the river.[32]

In the year 2000, the park expanded the existing viewing platform at Brooks Falls, creating a large complex with nearly 975 feet (300 m) of boardwalk with the intention of physically separating people and bears in this critical feeding area. Terry De-Bruyn and others took the opportunity to evaluate the effect such viewing structures had on bears.[33] They found that bears reduced their use of trails that passed under the boardwalk. One limitation of this work was that data collection only included 1 year before and after the installation of the boardwalk. This data limitation reduces our ability to draw inferences from it, as there could have been other factors unique to those years that influenced travel patterns near the falls. Another possibility is that if the boardwalk did impact bear movement as observed, as bears became habituated to the boardwalk in subsequent years, it may not have had the same impact on movement patterns.

In 2016, we began a new project to evaluate genetic connectivity throughout the park and examine the genetic health of Katmai's brown bear population. We found a high level of genetic diversity in all areas of the park, exceeding the average genetic diversity of brown bears in North America.[34] Intriguingly, we found some evidence of genetic isolation between interior areas of the park around Brooks Camp and the Katmai coast. This addressed a question the park had as to whether the bears were regularly crossing the Aleutian Range or if the relatively low resource availability across the top of the mountains and passes would result in more isolated groups. As some emigration is required to eliminate genetic isolation, this suggests that the bears in these areas have extremely high site fidelity and are not moving across the Aleutian Mountain Range very often. One possibility is that the genetic distinction is due to the formation of matrilineal assemblages at concentrated feeding areas. This has been documented in brown bears previously, where the local population is largely the offspring of a single female.[35]

A management concern has also been that high levels of visitation to Brooks Camp may exclude a portion of the population from the area and result in reduced genetic diversity. We found no evidence of reduced genetic diversity in Brooks Camp compared to other areas of the park, thus alleviating those concerns. We were also able to compare the genetic diversity at Brooks Camp to samples collected in 2006–2007 and did not detect any change in diversity over that time, though with the longer generation time of brown bears, we would not expect to detect any change over that period.[36] As high levels of visitation are a relatively recent addition to Brooks Camp, continuing to monitor for any impacts to the population is critical.

Another finding of the genetic research was evidence of a population bottleneck that had occurred within approximately the past 200 years.[37] While we do not have enough information to know with certainty the cause of this bottleneck, one likely possibility is the eruption of Novarupta in 1912. Following the eruption, there were reports of reduced salmon stocks in the fishery along the coast and additional contemporary reports of bears that were blinded by falling ash. Between direct mortality from the ashfall and loss of the primary food resource, it could reasonably be anticipated that this could have caused a temporary reduction in the bear population.

Effects of Human Use on Coastal Bears

Those familiar with the coast report that bear viewing did not exist prior to the Exxon *Valdez* oil spill in 1989. Ten years later, by 1998–1999, human visitation had started to increase, and Howard French examined the effects of the growing bear-viewing and photography industry on bears' spatial use patterns. The study at Hallo Bay revealed the behavioral effects of visitors on bears' seasonal movements. Bears congregate on coastal salt marshes in high densities in May, June, and July and then move to salmon streams for the remainder of the non-denning period. When people were present, bears exhibited increased vigilance.[38] Visitation at

this time was around 300 people per year. Between 2007 and 2009, Carissa Turner examined the effects of bear viewers again, though this time on a prominent salmon stream rather than at a salt marsh.[39] Time-lapse photography was used at Geographic Harbor to document when people and bears used the area and how human use influenced bear use of the stream. Turner found that the number of bears using the stream was high in the mornings and evenings and lowest during the middle of the day, when human visitation was highest. Models supported the idea that human presence was a significant factor in how bears used the stream.

From 2012 to the present, time-lapse work has been expanded to additional sites, including other salt-marsh and salmon stream sites with varying levels of human visitation. This work is ongoing but suggests that determining the effects of human visitors on bears may be more difficult than previously anticipated. Bears appear to show habitat-specific trends in daily use that may occur regardless of human visitation. Bears on salt marshes with no, low, and high human use seem to show increasing use throughout the day, even when people are present. Bears on salmon streams with no, low, and high hu-

▶ Although salmon and vegetation are bears' most common foods, some individuals are eating more marine-based food, such as clams, barnacles, flounder, seals, and sea otters (pictured here). (NPS/Deborah Kurtz)

man use seem to show morning and evening peaks in use regardless of the level of human use.[40] More work is needed to disentangle natural trends of use from those that change in response to visitors.

Diet and Habitat Use

In the period 2015–2017, we revisited earlier studies with new data collection aimed at understanding how diet and habitat use have changed through time.[41] We found that bears' diets were substantially different than they had been in 1989. In 1989, bears' assimilated diets were 62% salmon and 31% plant matter, but in 2015–2017 their diets were 27% salmon and 72% plant matter.[42] The number of bears observed at Hallo Bay also appears to have decreased 64% between 1998–1999 and 2015–2017—from 11 to 4 ± 4 bears annually.[43] Decreases in the abundance of chum and pink salmon,[44] warm-water ocean events, and changes in human visitation have all occurred during the time between studies and could be contributing to the changes we observed.

While studying diet, several researchers documented important linkages between the marine and terrestrial environments. Although vegetation and salmon seemed to provide the bulk of nutrients to bears on average, some individuals were found to consume less common foods, such as clams, barnacles, flounder, seals, and sea otters.[45]

A primary question for visitors and park staff has always been whether Brooks Falls bears in the interior of the park interact with bears on the coast. We were finally able to answer this question using habitat and genetics studies in the years 2015–2017. By examining GPS (Global Positioning System) collar data, we found that female bears collared on the coast remained on the coast and did not travel inland into other regions of the park.[46] Genetics work also showed that interior (Brooks Falls) bears have significantly different genetic signatures than females on the coast, suggesting that the 2 groups of bears do not regularly intermix.[47]

Aerial Surveys

Aerial surveys have been used to count bears congregating on the landscape starting in the 1970s with biologist Will Troyer and in regular use since 2013. Flying at low altitude (around 300–500 feet [90–150 m] above the ground), the pilot and an observer work together to spot bears at salmon streams and along shorelines. These surveys monitor the timing and use of salmon spawning streams by bears throughout Katmai National Park and Preserve and Aniakchak National Monument and Preserve.

Bear research conducted on the Katmai coast during the period 1989–1996 looked at the effects of the Exxon *Valdez* oil spill.[48] After finding no population-level effects of the oil spill on bears, analysis shifted to the population density of bears along the coast and elsewhere. As part of a larger study comparing brown bear densities across the state of Alaska, Katmai was found to contain the highest density of bears ever recorded (479 bears ≥2 years old/386 square miles [1,000 km²]).[49] Because animal density is often assumed to correlate with resource quality, it has been widely accepted that Katmai has superior resource quality for bears compared to other areas.

Spring aerial surveys were conducted in the years 2003–2005 to estimate bear density in all of Katmai National Park and Preserve. The estimated bear density in Katmai National Park and Preserve was 156 ± 21 bears/386 square miles (1,000 km²) (population estimate of 2,183 ± 379 bears).[50] A third study was conducted in the Katmai Preserve during 2009, but it wasn't published; it estimated bear density at 101 ± 18 bears/386 square miles (1,000 km²) (population estimate 127 ± 23 bears).[51]

The Katmai Preserve is of particular interest because both sport and subsistence bear hunting is allowed (and not allowed within the park boundary). Therefore, it is of management interest to know trends in bear abundance and whether hunting pressure is affecting the bear population. Research has begun to incorporate aerial surveys of salmon streams with density estimates to help better estimate trends in the bear population and stream use.

Visitor Experience of Bear Viewing

In 2017, we initiated a study of backcountry visitor experience, in collaboration with Kansas State University and Clemson University, to inform wilderness- and visitor-use management planning. In Katmai, close encounters with bears are common and guided visitor groups often find themselves in the path of approaching bears, but how close do visitors actually want to be to a bear? This study included evaluating visitors' bear-viewing experiences, including number of bears seen, proximity to bears, and level of perceived crowding.[52] Regardless of proximity, visitors were likely to report a high-quality experience if they saw a bear. Visitors expected to see bears at an average of about 25 yards (23 m) and experienced an average distance of about 22 yards (20 m). Visitors were more likely to report that it was unacceptable to be closer than 25 yards (23 m) from a bear, with about 10 yards (9 m) too close and about 220 yards (201 m) too far away. Visitors reported that viewing a bear at a distance of between 25 yards and 100 yards (23–91 m) was acceptable and within 25–50 yards (23–41 m) was the most acceptable.

Visitors saw slightly more bears per hour than they expected to see and considered seeing 3–11 bears acceptable. Those who saw at least 2 bears within an hour reported a high-quality experience. Visitors at Hallo Bay considered viewing bears at closer distances more acceptable than did visitors

Crowding is not desirable for either visitors or bears. Studying visitor experience and perception of crowding helps the park determine thresholds for visitor management. (NPS)

at other locations. Overall, visitors reported experiencing bear-viewing conditions that were acceptable and supported a high-quality experience.[53]

Crowding is not desirable for visitors, and it is also not good for bears. Increases in visitation, levels of crowding at bear-viewing sites, and visitor proximity to bears will need to be monitored through the lens of visitor experience as well as impacts to bears. This study found that most visitors reported low levels of crowding while bear viewing; however, on average they encountered more groups (5) than they expected (3.5).[54] A better understanding of visitor expectations and experiences allows us to determine the thresholds at which management action should be taken to manage the number of people at sites and determine when bear-viewing practices need to be modified if conditions change.

Bears and Wolves

Bears share their space on the coast with another top predator: the wolf. Wolves occupy much of the same habitat as bears as they search for inter-tidal resources, scavenge, fish, and hunt. Interactions between wolves and bears appear to range from indifferent to defensive to playful to mutualistic and even predatory.

Wolves have been seen sitting close to and fishing alongside bears as they watch for salmon swimming upstream. Bears and wolves often pass each other with no apparent reaction or engagement. However, sometimes they push boundaries and appear to test dominance. A wolf was seen circling and repeatedly charging a bear cub in a family group, with the cub charging back before the wolf continued down the beach. A bear walking along a salmon stream came upon a sleeping wolf on the stream bank and charged at it, pushing it away from the stream; then, they both stood briefly at a standstill until the bear moved on. A wolf at a known rendezvous site was lying down near the mouth of a creek when a bear slowly started to move in its direction; when it turned its back, the wolf quickly ran and bit the bear's ankle. This was perhaps to defend the area, as young wolf pups had been seen there earlier that day. A young wolf and a sub-adult bear were repeatedly seen taking turns chasing each other on the inter-tidal. On a different day, the wolf was seen crouching and sneaking up on the bear as it was sleeping, disrupting it from its nap and then chasing it. Wolves have also scavenged or possibly preyed on bear cubs, with reports of wolves carrying dead bear cubs and evidence of bear remains in numerous wolf scats. There are undoubtedly many untold bear and wolf stories from the Katmai coast. Though bears and wolves are competing for some of the same food sources, the abundance of resources on the Katmai coast allows both bears and wolves to thrive. Their relationship is sometimes even mutualistic, as wolves feed on

leftover carcasses of fish caught by bears and bears scavenge prey hunted by wolves.

Successful brown bear conservation on the Katmai coast will require an understanding of how human visitation affects bears. Multiple researchers have investigated and found effects of human influence on bears in coastal Katmai, and still others suggest that interpreting these studies may not be as straightforward as previously thought. More needs to be understood to allow us to develop effective management plans.

Brooks Camp presents a prime natural laboratory for examining the impacts of people on bear biology and behavior. While several observational studies have been done at Brooks and throughout Katmai, little to no work has examined the physiological or population-level effects on bears in areas of high human use. As Mark Ditmer and others documented sharp increases in heart rate despite little to no visible change in behavior in black bears approached by drones, it seems possible that the lack of visible response to aircraft, bear viewers, and anglers hides a more dramatic stress response

Wolves also occupy much of the brown bears' habitat. Mostly, they share the space with little reaction to each other, but some instances have been documented in which one has charged the other to test dominance. (NPS/Maurice Whalen, top; NPS/Matt Harrington, center; NPS/Kelsey Griffin, bottom)

among brown bears at Brooks Camp.[55] These possible physiological impacts, as well as how habituation affects them, should be investigated, as elevated stress can affect reproduction and life expectancy. Our genetic assessment of the park and preserve provides a baseline of genetic diversity and a comparison to other regions nearby. Continued monitoring of the diversity of these bears will provide insight into the health of the population and any impacts from increasing human presence. We believe future genetic monitoring should transition from micro-satellite data to the more sophisticated whole genome sequencing. This will improve the quality of the data, allow a more precise view of the genetic health of the populations, and allow us to explore more wide-ranging questions of the impacts of people on bears.

Little is known about the movement of bears seasonally away from the areas of high abundance along salmon streams and sedge meadows. As radio and GPS collars have not been used with this population since the 1970s, seasonal movement and denning locations are unknown. These data are critical to understand how the bear population in Brooks Camp and throughout the park and preserve may be impacted by management actions both within and outside the park. Bears frequently seen in Brooks Camp have been caught in wolf snares in the nearby community of King Salmon, and understanding whether bears managed within the park

are regularly feeding at the dump in town is critical for making important decisions regarding the safety of employees and visitors.

Understanding the impacts of climate change on this ecosystem is of vital concern for park management. As mentioned, the impacts on salmon have the potential to dramatically alter the entire flow of nutrients within the ecosystem. In addition to potential impacts to the salmon run, however, climate change may affect timing of food availability and den emergence in brown bears.

Across the Shelikof Strait on Kodiak Island, a change in the timing of the elderberry crop has modified the time when bears leave the salmon runs to seek berries.[56] Such timing changes may have a great impact on the future of both bear and human use of Katmai.

Salmon play an especially central role in bears' lives, yet little is known about Katmai's coastal salmon populations. No salmon-related studies have been done since bear research began on the coast in the late 1980s, and current data are limited to occasional aerial surveys conducted by the Alaska Department of Fish and Game in Kodiak. Research on the timing and abundance of coastal salmon runs would help us detect changes in the marine environment and the resulting impact on bears. At the same time, if vegetation continues to become increasingly important for bears, vegetation studies—including understanding the role of

salt marshes for bears—could also become increasingly valuable in predicting how bears might respond to changing conditions.

Similarly, successful brown bear conservation on the Katmai coast will require an understanding of how altered resource availability affects bears. As the environment continues to change, it will be important to continue research on coastal bear diet and the implications of dietary shifts. Bear populations respond to food availability, especially meat availability, with changes in cub production and population density.[57] Therefore, the recent changes to bear diets could have population implications that warrant better understanding. Was the change in diet and bear abundance observed during the years 2015–2017 a response to a short-term change in ocean conditions, or does it signal a new normal for bears? If bears must rely more heavily on vegetation and other resources in the absence of salmon, how does that impact their survivorship and productivity? What impact could such a substantial change in diet have on the bear population? Katmai's bear population has not been resurveyed

◄ Bear populations respond to food availability. This bear was observed earlier in the year than normal, when most bears are still in their dens, and was in better condition than most bears in early season. For bear conservation into the future, it will be important to understand changes in a diversity of food sources and how bears are changing with them. (NPS/Deborah Kurtz)

since 1990. To better speak to the health of Katmai's coastal bears, a better understanding of population trajectory would be incredibly valuable. Future research on harvest management should attempt to provide estimates of bear abundance along salmon streams, corrected for time of year and imperfect detection. This will allow managers to understand when and where bears will be using streams throughout the park and can allocate resources accordingly.

There is also a need to better understand additive impacts to bears, such as multiple years of low salmon numbers, ocean warming, climate change, ocean acidification, increased visitation, and other anthropogenic effects. Future brown bear research will need to factor in human dimensions and implications for park management as we strive to protect this iconic species and preserve resources for future generations.

NOTES

1. Dumond, *Archaeology on the Alaska Peninsula.*
2. Luehrmann, *Alutiiq Villages under Russian and U.S. Rule.*
3. Woodhouse-Beyer, "Artels and Identities."
4. Clemens and Norris, *Building in an Ashen Land.*
5. Ringsmuth, *At the Heart of Katmai.*
6. Ringsmuth, *At the Heart of Katmai.*
7. Ringsmuth, *At the Heart of Katmai.*
8. Shepherd and Frith, *Monitoring Visitor Use in the Southwest Alaska Network.*
9. Alaska Department of Fish and Game, *Anadromous Waters Catalog.*

10. Troyer and Faro, *Aerial Survey of Brown Bear Denning in the Katmai Area of Alaska.*
11. RAWS data 2011–2021.
12. Shepherd and Frith, *Monitoring Visitor Use in the Southwest Alaska Network*; French, *Effects of Bear Viewers and Photographers on Brown Bears.*
13. Fortin et al., "Impacts of Human Recreation on Brown Bears."
14. United States Department of the Interior, National Park Service, *Final Environmental Impact Statement.*
15. Piatt et al., "Extreme Mortality and Reproductive Failure of Common Murres."
16. National Park Service and Alaska Department of Fish and Game, *Best Practices for Viewing Bears on the West Side of Cook Inlet and the Katmai Coast.*
17. Piatt et al., "Extreme Mortality and Reproductive Failure of Common Murres."
18. Troyer, *Into Brown Bear Country.*
19. Troyer, *Distribution and Densities of Brown Bear on Various Streams in Katmai National Monument.*
20. Troyer, *Movements and Dispersal of Brown Bear at Brooks River, Alaska.*
21. Braaten and Gilbert, *Profile Analysis of Human-Bear Relationships in Katmai National Park and Preserve.*
22. Braaten and Gilbert, *Profile Analysis of Human-Bear Relationships in Katmai National Park and Preserve*; Olson and Gilbert, "Variable Impacts of People on Brown Bear Use of an Alaskan River."
23. Olson and Gilbert, "Variable Impacts of People on Brown Bear Use of an Alaskan River."
24. Olson, Squibb, and Gilbert, "Brown Bear Diurnal Activity and Human Use."
25. Olson, Squibb, and Gilbert, "Brown Bear Diurnal Activity and Human Use."
26. Olson, Gilbert, and Squibb, "The Effects of Increasing Human Activity on Brown Bear Use of an Alaskan River."
27. Smith, "Effects of Human Activity on Brown Bear Use of the Kulik River, Alaska"; Smith and Johnson, "Modeling the Effects of Human Activity on Katmai Brown Bears."
28. Smith, "Effects of Human Activity on Brown Bear Use of the Kulik River, Alaska."
29. Smith and Johnson, "Modeling the Effects of Human Activity on Katmai Brown Bears."
30. Smith, "Effects of Human Activity on Brown Bear Use of the Kulik River, Alaska."
31. Fortin et al., "Impacts of Human Recreation on Brown Bears."
32. Skora, "Population Dynamics of Brown Bears along Brooks River in Katmai National Park, Alaska."
33. DeBruyn et al., "Brown Bear Response to Elevated Viewing Structures at Brooks River, Alaska."
34. Saxton, "Investigating Population Genetics."
35. Støen et al., "Kin-Related Spatial Structure in Brown Bears."
36. Saxton, "Investigating Population Genetics."
37. Saxton, "Investigating Population Genetics."
38. French, *Effects of Bear Viewers and Photographers on Brown Bears.*
39. Turner, "Determining the Effectiveness of Park Management Strategies."
40. Griffin, *Spatio-Temporal Distribution of Coastal Brown Bears and Visitors*; Erlenbach, Griffin, and Robbins, *The Need for Habitat-Specific Management of Bears and Bear Viewing.*
41. Erlenbach, "Nutritional and Landscape Ecology of Brown Bears."
42. Sellers et al., *Population Dynamics of a Naturally Regulated Brown Bear Population*; Hilderbrand et al., "The Importance of Meat."
43. Smith and Partridge, "Dynamics of Intertidal Foraging by Coastal Brown Bears in Southwestern Alaska."
44. Malick and Cox, "Regional-Scale Declines in Productivity of Pink and Chum Salmon Stocks."
45. Monson et al., "Brown Bear–Sea Otter Interactions along the Katmai Coast"; Sellers et al., *Population Dynamics of a Naturally Regulated Brown Bear Population*; Smith and Partridge, "Dynamics of Intertidal Foraging by Coastal Brown Bears in Southwestern Alaska."
46. Sellers et al., *Population Dynamics of a Naturally Regulated Brown Bear Population.*
47. Saxton, "Investigating Population Genetics."
48. Sellers et al., *Population Dynamics of a Naturally Regulated Brown Bear Population.*
49. Miller et al., "Brown and Black Bear Density Estimation in Alaska."

50. Olson and Putera, *Refining Monitoring Protocols to Survey Brown Bear Populations*.
51. Loveless et al., "Population Assessment of Brown Bears in Katmai National Preserve, Alaska."
52. Brownlee et al., *Evaluation of the Bear Viewing Experience and Associated Thresholds*.
53. Brownlee et al., *Evaluation of the Bear Viewing Experience and Associated Thresholds*.
54. Brownlee et al., *Evaluation of the Bear Viewing Experience and Associated Thresholds*.
55. Ditmer et al., "Bears Show a Physiological but Limited Behavioral Response to Unmanned Aerial Vehicles."
56. Deacy et al., "Phenological Synchronization Disrupts Trophic Interactions."
57. Hilderbrand et al., "Importance of Meat."

Katmai is one of the best places to study bears and there is still much to learn through future research. (NPS/Kelsey Griffin)

The coastline of Lake Clark National Park and Preserve has been known for congregations of brown bears since at least the early 1920s. (NPS/Kara Lewandowski)

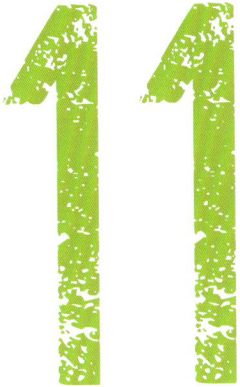

11

LAKE CLARK

DIVERSE LIFE HISTORIES AMID MYRIAD PRESSURES

Buck Mangipane, Lindsey Mangipane, Karen Evanoff, and John Branson

HISTORY

For thousands of years, bears and people have lived in what would become Lake Clark National Park and Preserve (Lake Clark). Revered by Dena'ina Athabascans, their coexistence on these lands for many generations provided ample opportunity for observations, and observations led to knowledge. This close relationship and connection with the natural world created an intimacy unknown to most, cultivated through a life immersed in nature where one learned to sense its deeper intelligence. This is called *Q'et ni' yi* in the Dena'ina language, meaning "it is speaking to us," referring to the natural world.

Highly respected, Dena'ina used the indirect term *big animal* rather than the direct name of bears to prevent any offense or disregard to the big animal. In turn, good things would come to them in return. To show respect when a bear was taken, it was important to cut out the eyes and bury them so the spirit could not see what was happening. Another reason for kindness and high regard for big animals is their close resemblance to human beings when standing.

https://doi.org/10.5876/9781646427116.c011

Taken bear provided much to the Dena'ina. Fur was used to make blankets, parkas, and mittens. Intestines and stomach linings were dried and used to make rain jackets and sacks and could be used to line birch baskets to store fat. The meat was good eating, and the fat was rendered for oil and did not freeze solid during the cold winters. Given the scarcity of moose and caribou in the past, bear provided significant meat.

While Russian and American exploration in the Lake Clark region began earlier, we start to find a few references to bears as we enter the early twentieth century. Upon exploring the region, Frederick K. Vreeland wrote to Dr. C. Hart Merriam in December 1921, "This northwest shore of Cook Inlet is said to be a great place for bear . . . I was quite shocked to see how the brown bears have been destroyed along the southwest coast [Cook Inlet?]. They have very few friends in that region."[1] It is interesting to note that impacts to bears were already noted at this early date. Along Lake Clark we find a more favorable description of the bear population in a letter from Otis M. "Doc" Dutton and J. E. Kackley of Tanalian Point, Alaska, to Colonel A. J. Macnab in 1924, "There is lots of sheep now & brown bear . . . We are going to have a bear hunt this spring up Little Lake Clark or up Brooks Lake [Kontrashibuna Lake] we havn't [sic] decided as yet. 10 ft. prime skins will sell at the cannery for a good price."[2] Many bear hides from this region were used for warmth by crews manning the double-ender sailboats used in commercial salmon fishing in the early days of Bristol Bay.

Lake Clark was established on December 2, 1980, under the Alaska National Interest Lands Conservation Act (ANILCA). Located in southwest Alaska, the approximately 4 million acres of land is a microcosm of Alaska. This diversity is reflected in its purpose, which includes maintaining unimpaired scenic beauty, protecting the watershed necessary for the perpetuation of the red salmon fishery in Bristol Bay, and protecting habitat for and populations of fish and wildlife, among others. Specifically mentioned among the suite of wildlife inhabiting these lands are brown/grizzly bears.

ENVIRONMENT

Located in southwest Alaska, Lake Clark covers over 4 million acres of land. Elevations range from sea level along the Cook Inlet coastline to Mount Redoubt's 10,197-foot (3,108-m) summit. Two extensive mountain ranges, the Alaska Range from the north and Aleutian Range from the south, meet in the park. The resulting intersection is a jagged array of mountains and glaciers known as the Chigmit Mountains. It is within this range that the park's 2 active volcanoes, Mount Redoubt and Mount Iliamna, are found. On the east side of the Chigmits, higher elevations consist of steep, barren slopes that give way to tundra and dense alder that descend to the coast. As the landscape flattens, spruce forests take shape. In the valleys, broad wetlands are bisected by rivers and streams lined with cottonwoods, alders, and willows. At the mouths of some bays, expansive salt-marsh meadows form by salt-tolerant plants that survive the daily inundation of the tides. West of the Chigmits are broad expanses of alpine tundra that transition to shrublands of willow and alder as elevation decreases. In the northern portion of the park, spruce forests dominate the valley bottoms, whereas mixed forests of spruce and birch cover much of the park's southern portion. Interspersed within these forests, small ponds and wetlands occupy low-lying areas.

The climate of Lake Clark falls within 2 distinct climatic regions: the damp coast and the drier interior. Precipitation on the coast averages 40 inches (101 cm) to 80 inches (203 cm) annually, with foggy, wet conditions common. In the interior, precipitation averages 17 inches (43 cm) to 26 inches (66 cm) annually. Winter temperatures can drop to below –40°F (–40°C), with an average January high temperature of 22°F (–6°C). Frost and snow can occur any time but are most common from September to early June. The

Lake Clark National Park and Preserve encompasses more than 4 million acres in southwest Alaska. It includes 2 extensive mountain ranges—the western edge of the Alaska Range in the north and the volcanic Aleutian Range in the south.

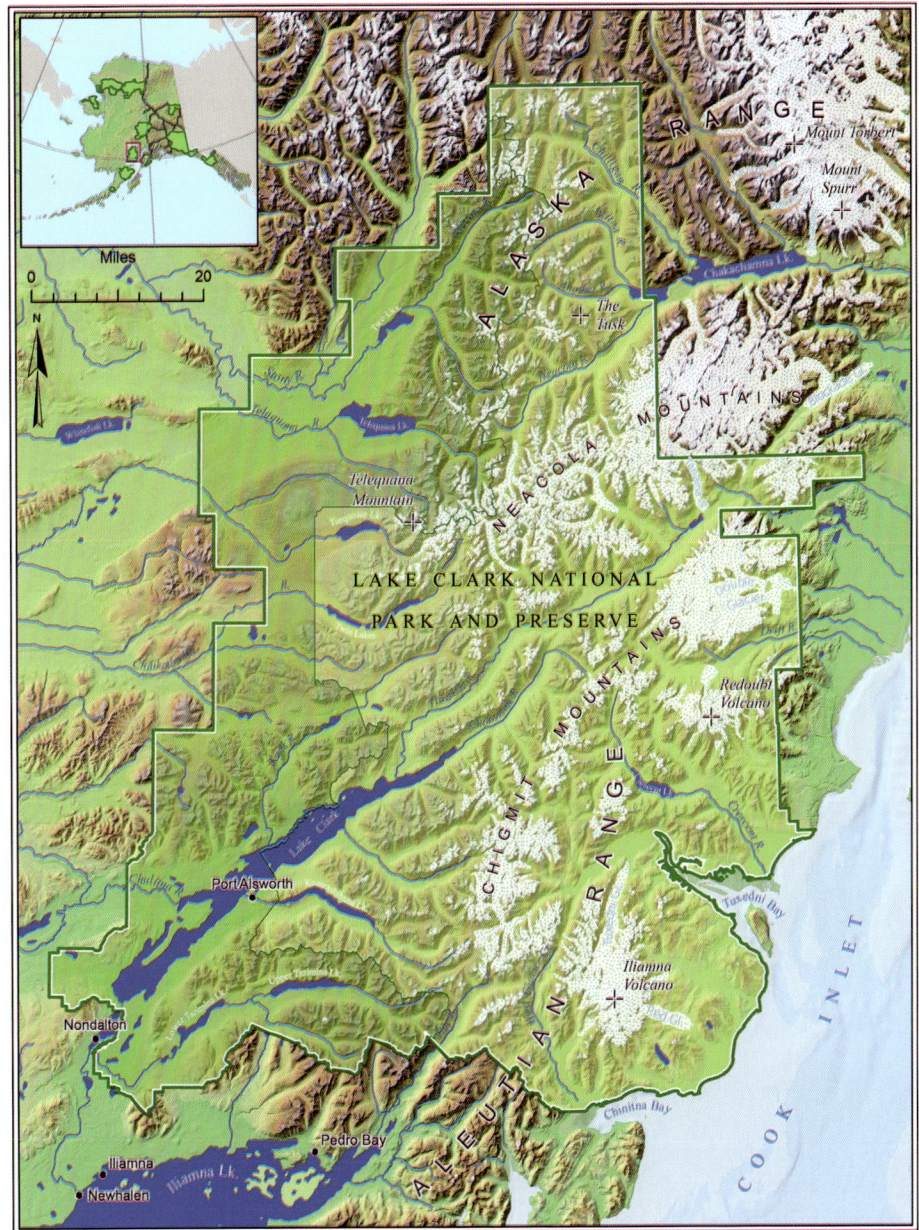

average temperature in July is 68°F (20°C), with highs over 80°F (27°C) occurring infrequently. During the short summers of dry years, fires occasionally burn west of the Chigmit Mountains.

The park's mountainous spine distributes its water among multiple watersheds. Coastal draining rivers such as the Neacola, Crescent, Tuxedni, and Johnson Rivers generally flow east and end in Cook Inlet. West of the Chigmits, a series of large lakes extends from north to south. The major lakes in the park's interior (Two Lakes, Telaquana Lake, Turquoise Lake, Twin Lakes, and Lake Clark) sit at the headwaters of 3 major river systems in southwest Alaska. In the northernmost portions of the park, the Stony, Necons, and Telaquana rivers drain west into the Kuskokwim River, terminating in the Bering Sea. Further south and emerging from Turquoise Lake and Twin Lakes, the Mulchatna and Chilikadrotna rivers, respectively, drain west before joining outside the park to enter the Nushagak River, which flows into Bristol Bay near Dillingham. Continuing south, the Tlikakila, Chokotonk, Kijik, Chulitna, and Tanalian rivers flow into Lake Clark, which drains to the south by way of the New Halen River into Lake Iliamna. There it continues into Bristol Bay near Naknek via the Kvichak River. These pristine waters support a diverse fish assemblage. Resident whitefish, Arctic grayling, northern pike, lake trout, rainbow trout, and Dolly Varden are found in many waters throughout the year.

All 5 species of salmon spawn within the waters of the park, with sockeye salmon the most numerous. Large numbers of sockeye salmon ascend the park's rivers, with significant numbers spawning in Telaquana Lake, Lake Clark, and Crescent Lake, providing an important food resource for bears.

Lake Clark is home to a full complement of native wildlife. Among these, Arctic ground squirrels, caribou, Dall's sheep, and moose are prey for brown bears. Moose are broadly distributed within the park, with calving from mid- to late May to mid-June. Newborn moose are preyed upon by brown bears throughout their range, but calves quickly become large enough to evade bears, lessening predation impacts. Caribou inhabit the western portion of the park. Their numbers and distribution vary greatly. Similar to moose, they are typically most vulnerable to bears as very young calves. Dall's sheep reach the southwest extent of their geographic range in the park and use steep, rugged terrain during periods when bears are active, which tends to limit their susceptibility to predation. As bears emerge from their dens, they can also find and take advantage of large mammals weakened or killed by the long, harsh winter. Throughout the park,

▶ The mountainous park has multiple watersheds and many lakes and rivers. All 5 species of salmon spawn in park waters, providing important food for bears. (NPS/Ken Miller)

vegetation plays a large role in bears' diets. This is most notable on the coast, where large numbers of bears congregate in salt-marsh communities during June and July to feed on sedges and other nutritious plant species. In late summer and into fall, a variety of berries becomes an important food source for bears. This rich diversity of resources helps support the moderate to high densities of bears found throughout the park.

MANAGEMENT CHALLENGES

Like many parks, Lake Clark faces a wide range of management challenges. Similar to many units of the National Park System, visitation has become an emerging and challenging issue in recent years. Between 2010 and 2019, the park saw visitor-use days (each day a visitor is at the park engaging in some activity) increase from 4,000 days to over 19,000 days a year. During this period, the park also saw visitors' primary activities change, with bear viewing becoming the most popular activity, surpassing sport fishing. During that 10-year period, bear viewing increased in popularity from just over 1,500 annual visitor-use days to nearly 8,000.[3] While the number of visitors viewing bears may

Moose are distributed broadly across the park. Although moose calves are preyed upon by brown bears, the calves quickly become large enough to evade them. (NPS/Kim Arthur)

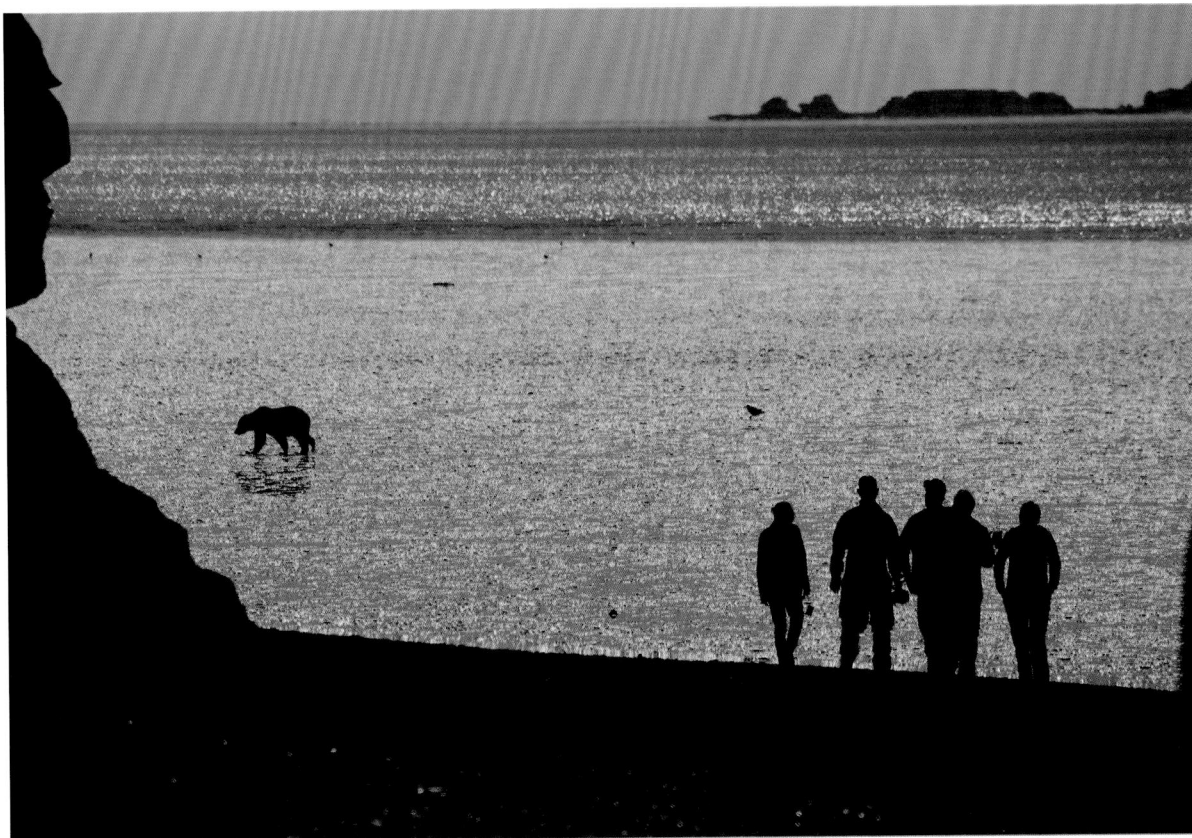

Between 2010 and 2019, visitors' bear-viewing days increased from 4,000 to more than 19,000 a year, nearly all of which occurred between June and August at just 3 sites: Crescent Lake, Silver Salmon Creek, and Chinitna Bay. (NPS/Jim Pfeiffenberger)

seem low, the activity is concentrated in space and time. Crescent Lake, Silver Salmon Creek, and Chinitna Bay receive nearly all the visitation during the short time period June through August. Increased visitation may disturb bears to the point that they will potentially avoid these areas. Given the high value of the salt-marsh and salmon resources these areas contain, this could have impacts on individual bears and, ultimately, bear populations. Given the likelihood of continued increases in visitation, park managers will need to understand what levels of visitation would negatively affect visitor experience, bear behavior, and bear populations.

Lake Clark is large and remote and has historically had limited staff. These circumstances can mean that many visitors never encounter park employees

during their visits, which can reduce their ability to get information about how to behave responsibly in bear habitat. While reports of human-bear conflicts are low, increased visitation throughout the park has also increased the potential for negative interactions between people and bears. In high-visitation areas, the park has implemented a variety of measures to reduce human-bear interactions, primarily by minimizing the availability of items that attract bears (i.e., attractants management). This includes regulations on the coast that relate to proper food storage, fish handling and storage requirements, and required use of electric fences—all of which help decrease the likelihood of negative human-bear interactions. While effective, increased visitation, noncompliance with attractant management, and limited staffing are factors that allow conflicts between people and bears to persist.

Both sport and subsistence hunting of brown bears occur in Lake Clark. Hunting of brown bears on the Alaska Peninsula is world-renowned, with the Lake Clark Preserve providing opportunities for sport hunting of this species. Subsistence hunting of brown bears occurs in the interior of Lake Clark and provides a culturally important resource for residents of Iliamna, Newhalen, Nondalton, Pedro Bay, and Port Alsworth.

◄ Playful cubs are always a favorite attraction for visitors. (NPS/Kara Lewandowski)

Hunting management varies greatly across Game Management Units. Seasons and bag limits range from conservative, where management is specifically for trophy brown bears with hunts open every other year (the fall of odd-numbered years and the spring of even-numbered years) and individual hunters have a harvest limit of 1 bear every 4 years, to areas with an annual season with a 2 brown bears per year harvest limit. Given the harvest potential of Lake Clark, monitoring population size and trend is a critical data need to help balance bear harvest with mandates to protect bear populations.

In 1980, when it was established in Section 201(7) of ANILCA, the United States Congress significantly expanded the boundaries of the original park proposal to the shores of Cook Inlet. Within this expanded area, the law specifically excluded privately owned lands. Those private lands included the Johnson Tract, owned by Cook Inlet Region, Inc. (CIRI), long known for its mineral potential. During the 1971 Alaska Native Claims Settlement Act (ANCSA) Native lands selection process, CIRI was unable to make its full land selection because most of the land in the Cook Inlet region was already under private, federal, municipal, or state ownership. Ultimately, CIRI's land selections were fulfilled by the Cook Inlet Land Exchange, which was passed by Congress and signed by President Gerald Ford in 1976 (Public Law 94-203 and Public Law 94-456, as amendments to ANCSA). As part of the land exchange agreement, CIRI received the 20,942-acre Johnson Tract, on the west side of Cook Inlet at the headwaters of the Johnson River. The Johnson Tract consists of 2 smaller parcels: the 11,342-acre South Tract, where CIRI received both surface and subsurface estates, and the 9,600-acre North Tract, where CIRI received the subsurface estate. Use of the South Tract by CIRI is restricted for purposes of mining and mineral extraction; CIRI's surface use of the North Tract is guaranteed for the purposes of access, exploration, and extraction of the mineral estate. In addition, to allow access to minerals from the remote location, a mutually agreed upon road and port site on Cook Inlet is stipulated. Gold and other mineral assets were discovered in 1982 by the Anaconda Company. During the 1980s and early 1990s, the deposit continued to be explored until a combination of factors caused efforts to cease. In 2019, JT Mining, Incorporated (JTMI), an Alaska-based company and wholly owned subsidiary of HighGold Mining, Inc., entered into a lease agreement with CIRI. JTMI assigned its rights under the lease to HighGold. The lease grants JTMI rights of access and the right to explore for and develop the Johnson Tract's minerals. This renewed exploration began in 2019 and continues at this time.[4]

In addition to potential development inside park boundaries, external development adjacent to the park has been proposed. Oil and gas lease sales on state and federal lands in Cook Inlet adjacent to the park occur regularly. To the southwest is the proposed Pebble Project, considered one of the world's largest undeveloped copper and gold resources. The project's owners, Northern Dynasty Minerals, Ltd., continue to work to obtain the permitting needed for development. The deposit sits within 2 watersheds with headwaters that begin in the park, and project infrastructure could include transportation routes adjacent to the southern boundary.

RESEARCH AND MONITORING

Historically, bears in Lake Clark have received limited research attention. We know of no scientific publications on brown bears from Lake Clark prior to the late 1990s. Because of management concerns, research has been initiated on bear demographics and ecology.

Population Monitoring

The protection of brown bear populations and their habitat is specified as a purpose in the establishing legislation for Lake Clark National Park and Preserve. ANILCA also provided opportunities for subsistence hunting by local residents throughout the park and sport hunting in the preserve. The Alaska Peninsula, with Lake Clark at its northern end, is world-renowned as a destination for trophy bears, drawing people from around the world to pursue

Wildlife biologists Lindsey and Buck Mangipane collect data on a bear while it is tranquilized for collaring. (NPS)

these animals. The large number of bears in the park, in seasonally accessible and visible habitat, has also led to large increases in bear viewing over the past decades. As park managers look to balance protecting bear populations with allowing hunting and bear-viewing opportunities, monitoring bear populations becomes a critical data need to make management decisions and inform harvest strategies.

In the late 1990s and early 2000s, we began to work on estimating the populations of bears in the park. Given the generally solitary nature and wide-ranging movements of brown bears, their populations are difficult to monitor. While a variety of techniques had been used previously to estimate bear density and population estimates, we chose an aerial survey method developed by the Alaska Department of Fish and Game (ADF&G) because it can work at large spatial scales, does not require radio-collared animals, and is relatively cost-effective.[5] Working with ADF&G, initial surveys focused on the Game Management Unit (GMU) 9B section of the interior of Lake Clark. Coastal surveys were conducted in 2003 and covered all of the park that falls within GMU 9A.

Following up on the bear survey, completed in May 2003, park staff began to consider additional ways to assess the bear population on the park's coast. While the completed survey promised robust data that would have high value, it required a rather large commitment of time and resources.

Those commitments made surveys of that scale something that would occur infrequently, possibly every 5 to 10 years. Supplementing those data with data collected annually and that would be able to serve as a population index was the goal of a new effort. We settled on an aerial index count of the coastline, with a focus on the salt-marsh meadows (Chinitna Bay, Shelter Creek, Silver Salmon Creek, and Tuxedni Bay) during late June and early July. We chose this region because the salt-marsh meadows provide seasonally important food resources that concentrate brown bears, and the bears that use these areas are easily visible from an aircraft. We established a goal of completing 2 index counts each summer in late June to early July, as bear numbers can vary with the greening and senescing of the meadows and that process varies across years. Between 2003 and 2022, we completed 36 index counts.

Building on the coastal population–level survey in 2003, we completed an additional survey using the same methods in 2010. After that survey, we evaluated the methodology and incorporated a modified conventional distance sampling approach (described by Josh Schmidt and colleagues) in the subsequent 2019 survey.[6] At this time, we discussed how the 2 population datasets could be integrated. Given that each approach had limitations, we hoped an integrated modeling approach would improve estimators of abundance and trend. Schmidt and others presented the integrated data incorporating the robust surveys (reanalyzed 2003 and 2010 data and 2019 survey data) with the salt-marsh index counts (2003–2019).[7] These data estimated the density of brown bears within our study area at 38–54 adults/1,000 km^2 during 2003–2019; that abundance increased at a rate of approximately 1.4%/year.

Ecology of Brown Bears in Lake Clark's Interior

Prior to 2014, we knew very little about bears in the interior of Lake Clark. With mounting data needs related to the potential development of Pebble Mine adjacent to the park, it became clear that research was needed to learn more about this unique segment of the brown bear population. Although brown bears are one of the most studied species in North America, there had never been a study focused on bears in the park's interior. In well-studied systems, such as the Greater Yellowstone Ecosystem and northwest Montana, biologists have

Brown bears depend on salt marshes at critical times of the year before other foods are available. As a warming climate causes sedges to green up and senesce earlier, it could have a large impact on bears. (NPS/Jim Pfeiffenberger)

Lake Clark has one of the most prolific salmon fisheries in the world, providing bears with a high-nutrient food source. (Mark Johnson)

built a large body of scientific literature that they continue to expand on to better understand complex aspects of bear ecology, behavior, and population dynamics. In Lake Clark, we had a blank slate. As biologists, this was the opportunity of a lifetime, and we were eager to learn as much as we could about these bears.

Although bears in the interior of Lake Clark don't have access to coastal food resources, they do have more food available to them than do bears in other interior regions of Alaska, such as the Brooks Range. The mountains of Lake Clark's interior are the headwaters of Bristol Bay, one of the most prolific salmon fisheries in the world. Each fall, the interior receives a significant influx of anadromous salmon as they return to their home waters to spawn, making Lake Clark a unique "middle ground" of food availability. This was particularly exciting for us, and we were eager to learn how bears make a living in this distinct system. Given the lack of basic information about brown bears in Lake Clark's interior, we started with 2 basic questions: (1) what are brown bears in the interior eating and (2) how do bears use the landscape to meet their resource requirements?

To answer these questions, we captured and radio collared brown bears in the interior portions of the park from 2014 through 2017. At the time of capture, we weighed bears, measured their proportion of body fat, took various physical measurements, and collected biological samples such as hair and blood, which could be used to evaluate diet. We also fitted bears with radio collars that collected a GPS (Global Positioning System) location every 90 minutes. In total, we captured 49 individual bears between October 2014 and May 2017.

When we first started to receive downloads of the location data from the radio collars, we were very surprised by what we saw. Some bears demonstrated movement patterns typical of what we expected, generally sticking to discreet areas composed of a few mountain valleys. We also observed bears on the opposite end of the spectrum that emerged from their dens and immediately headed more than 100 miles (160 km) west to areas where spawning salmon first enter the system. These bears then followed the salmon as they traveled east back toward the mountainous regions of the park and remained in proximity to anadromous streams until they entered their dens in the fall. Some of these bears followed the same movement path multiple years in a row. The dichotomy between space-use strategies was striking and highlighted the variety of approaches brown bears can use to make a living.

There is a large body of literature evaluating bears' nutritional ecology. Brown bears are generalist omnivores, meaning they eat a wide variety of plants and animals. Although bears are generalists at the population level, individuals can have more specialized diets, resulting in generalist populations made up of specialist individuals. For example, bears in Denali National Park and Preserve and Denali State Park were found to have highly variable diets, ranging from predominantly herbivorous to diets consisting of large amounts of salmon.[8] Foraging strategies among brown bears can largely be based on body size.[9] Smaller bears have lower daily energy requirements, which can make foraging on spatially dispersed foods such as berries more energetically feasible than it would be for larger bears that require more energy.[10] However, being big has its advantages. Large bears are generally more dominant and can have increased access to food and mates. Therefore, there are tradeoffs in foraging strategies among bears: stay small and have the ability to forage more efficiently, or benefit from being big while enduring the energetic costs of maintaining your large stature.

Body fat is an important indicator of bear health, with increased fat associated with increased reproduction and successful denning. Therefore, foraging behaviors that promote optimal fat gain are extremely beneficial to bears. Although the optimal diet is different for large and small bears, studies of captive bears have found that the most biochemically efficient diet for small-bodied bears is a mixed diet of ~ 24% protein-rich meat and ~76% carbohydrate-high fruits.[11] In late summer–early fall, bears enter a period called hyperphagia, in which they eat large amounts of food to pack on pounds for the long denning season ahead. In Lake Clark, the hyperphagic period overlaps with the time salmon enter the system and berries become ripe. Therefore, because the majority of body fat is gained during the hyperphagic period, we chose this time period to evaluate brown bear diet. As bears consume various foods, carbon and nitrogen atoms from the foods they eat are assimilated by the body. We are then able to compare the carbon and nitrogen values of the hair samples we collected at the time of capture with the values identified for bear foods, such as salmon, terrestrial meat, or vegetation, to determine the proportion of each food that makes up the bears' diet.

We found that the autumn diet of brown bears in Lake Clark's interior was highly variable, ranging from 40% to 87% meat (salmon and terrestrial meat combined), with the average diet 70% meat.[12] This was 46% greater than the expected optimal diet proposed in previous studies on captive bears, suggesting that "optimal" for wild bears may be more complex than that of captive bears. Furthermore, when salmon are abundant and available, bears may consume this resource under an energy-maximizing strategy rather than select for a nutritionally optimal diet. We also found that despite variable diets, bears were able to achieve similar levels of body fat, highlighting their ability to demonstrate a high level of plasticity, or the ability to use different strategies to achieve similar biological outcomes (see chapter 12 for more information on behavioral plasticity in bears). Body mass, however, was influenced by the percentage of meat in the diet, with larger bears consuming more meat than smaller bears. This is consistent with previous findings and suggests that most bears in Lake Clark's interior are likely using an energy-maximizing strategy during hyperphagia; however, smaller bears are still foraging closer to the nutritionally optimal diet than are larger bears.

Foraging decisions can largely drive how bears use the landscape, but other factors such as availability of cover and quality of resources can also contribute to space-use decisions. Under optimal foraging theory, animals optimize the use of resources while minimizing the amount of energy used to acquire them.[13] The distribution and seasonal availability of resources can markedly impact space use, with the area an animal uses smaller when resources are concentrated in space or time. Given what we learned about the diverse diets of bears in the interior of Lake Clark, we wanted to evaluate how home range size was influenced by seasonal availability and spatial distribution of resources and if reproductive status (solitary female versus female with cubs), sex, or body size influenced the way bears use the landscape. To do this, we evaluated bear movement data obtained from the radio collars and compared seasonal home ranges with various habitat characteristics, such as distance to salmon streams, availability of berry habitat, and measures of habitat diversity and richness.

We found that the spatial distribution of resources had the largest impact on brown bear home range size.[14] Bears had larger home ranges when landscape homogeneity increased, likely to gain access to a diversity of food resources. This pattern was observed among all sex and reproductive classes and was not influenced by the size of the bear. These findings are consistent with our findings related to bear diets and highlight the use of a diversity of food resources.

Wildlife biologist Dave Gustine monitors the heart and respiration rates of a sedated and collared bear. (NPS)

Although space use during active months is critically important to understanding brown bear ecology, it is equally important to understand the characteristics of winter dormancy. Winter denning is an important part of a bear's life, and it allows bears to reduce their metabolic activity and conserve energy during the harsh winter months when food resources are limited. Dens provide secure locations for bears when they enter this state of reduced activity, and they provide thermal insulation from extreme cold. In addition, dens provide a secure location for females to give birth to cubs. Through our research, we located 70 den sites (19 male and 51 female) from 40 individual bears (some bears were collared for multiple denning seasons).

Den-site selection can be based on a variety of factors, including elevation, steepness of the slope, aspect, proximity to food resources, and potential risk from other bears. Female bears have been known to alter their den-site selection to avoid adult males who sometimes prey on denning females or unrelated offspring.[15] Because of this, females have been known to den at higher elevations to spatially separate themselves from males, and they spend longer periods of time in the den to temporally avoid males. Based on these previous findings, we predicted that male and female bears would select for different den-site characteristics and have differences in the timing of den entrance and emergence.

We found that slope was the most important factor influencing den-site selection for male bears.[16] Moderate slopes with a grade of 20°–40° were selected for, and have previously been found to provide, the highest level of structural stability and drainage for excavated dens.[17] Males also selected den sites that were at lower elevations in proximity to habitats that likely provided productive fall foods, including blueberries and salmon. Lake Clark's interior has one of the latest sockeye salmon runs in Alaska, and fish are available to bears through late October. By denning at lower elevations, males are able to forage in productive habitats and expend minimal energy searching for a den site. In addition, winter-killed ungulates and vulnerable calves generally become available earlier at lower elevations, providing potential spring foraging opportunities following den emergence. Similar patterns were observed in Denali National Park and Preserve, where male bears selected den sites that overlapped with spring caribou calving areas.[18] Elevation was the largest driver of den-site selection for female bears. Females had the highest probability of denning at elevations near 2,940 feet (896 m; range = 328–4,920 feet [100–1,500 m]), whereas males on average denned at 2,500 feet (762 m). Slope was also an important factor for females; like males, they selected for slopes that would provide strong structural stability. To a lesser degree, selection of southerly aspects was also a contributing factor for female den-site selection.

On average, female bears entered their dens on October 20 and males followed, entering dens on October 28.[19] Similar patterns were observed for spring den emergence, where female bears exited dens 7 days later than males (May 8 and May 1 for females and males, respectively). The 8-day difference in den entrance dates between male and female bears was statistically significant and likely represents a tactic used by females to temporally avoid male bears. By entering dens earlier, females are able to travel through lower-elevation denning habitat used by males before they are present to prevent physical encounters and avoid leaving scent trails that could lead male bears to female den sites. Although female bears emerged from dens on average 7 days later than males, the difference was not significantly different. These findings are in contrast to previous studies evaluating brown bear denning chronology, which have found that females emerge from dens significantly later than males to avoid exposing cubs to potentially infanticidal males. However, because females in the Lake Clark interior den at higher elevations than males, they may be able to travel to lower elevations (where foraging opportunities become available sooner in the spring) behind males and still effectively avoid them.

Our research efforts provided some of the first insights into the ecology of brown bears in the park's interior and established valuable baseline information about the population. Although Lake Clark is largely unaffected by human activity, the potential for increased human activities, such as resource extraction, still exists just outside the park's boundary. Through our work, we were able to document bear movements both inside and outside the park's boundaries, learn how bears are using the landscape in relation to resource selection, evaluate seasonal diets, and document important denning habitat. As the environment changes due to climate change and potential development, having baseline information is vital for assessing change and developing mitigation measures to promote bear conservation.

Ongoing and Future Research

The coastal areas of Lake Clark support a high density of brown bears that provides ample opportunities for visitors to view high numbers of bears with relatively easy access from Alaska's largest population centers. This unique situation has made bear viewing the most popular activity for park visitors. Bear viewing has also been largely responsible for the park's significant rise in visitation, increasing from just over 1,500 visitor-use days in

2010 to nearly 8,000 in 2019.[20] Increased human presence in critical brown bear foraging areas can influence behavior,[21] including space use, temporal activity patterns, and productivity.[22] Bear behavioral changes may be accompanied by physiological responses that indicate relative levels of stress bears are experiencing. In the summer of 2022, we began a project to assess short-term stress levels in bears by measuring fecal glucocorticoid metabolite (FGM) concentrations as well as bioactive thyroid hormone (triiodothyronine [T3]) derived from non-invasively collected brown bear fecal samples at 3 sites on the coast: Chinitna Bay, Shelter Creek, and Silver Salmon Creek. These sites provide varying levels of human visitation and allow a variety of bear-viewing methods; thus, they are a means of assessing whether a relationship exists between human visitation and short-term stress in brown bears. We expect results from the fieldwork completed in 2022 to be published in 2025 and will evaluate whether they can help evaluate current and guide future management.

In 2021, an interdisciplinary team of researchers began a project to document the intricate connections of the nearshore food web, along the park's coast. The team seeks to identify linkages among ocean currents, clams, sedges, salmon, sea otters, and bears. This work will provide a baseline for the Lake Clark nearshore community and an opportunity for comparison to recent work completed in nearby Katmai National Park and Preserve. Work under way or completed has included clam surveys of Chinitna Bay and Silver Salmon Creek, sea otter surveys, foraging observations, and habitat mapping, as well as the first radio collaring of brown and black bears on the Lake Clark coast. The data from this work will help predict cascading effects from sea otter occupation and will help prepare future management for a changing coastal ecosystem.

Sea otters have well-described impacts on intertidal and sub-tidal marine communities. Otters have not occurred along the park's coast for more than a century due to the maritime fur trade, allowing razor clams to flourish in their absence. At present, substantial razor clam beds provide abundant prey for coastal bears. As sea otters expand their range north along the coast, razor clam densities are expected to decrease due to sea otter predation. That reduction could force bears to shift their diets and habitat use as razor clams decline, potentially affecting the distribution and abundance of bears and bear-viewing opportunities for visitors. Because of the unique history of sea otter extirpation and impending reoccupation, as well as the presence of razor clams and bears, the coast provides a unique opportunity to begin to study the cascading effects of sea otter reoccupation on the nearshore habitat and the marine-terrestrial food web connections among sea otters, bears, and humans.

FUTURE NEEDS

As visitation continues to grow in Lake Clark National Park and Preserve, it has become increasingly important to understand the effects of human activity on brown bears. As more visitors seek bear-viewing opportunities in the park, brown bear disturbance could increase and displace bears from important habitats. While displacement is observable, underlying stress responses in bears may occur with higher human presence and go largely unnoticed. Elevated stress levels could affect bear health, having long-term effects on individuals and, ultimately, populations. Research on potential impacts of bear viewing combined with results from ongoing work assessing stress response of bears will be essential for park managers as they evaluate options for managing this popular and growing activity.

A near-term need for bear management and research is collecting data to inform where future development inside the park's boundaries will occur. As resource exploration on the Johnson Tract moves forward, the need to determine the location of an access road and port becomes more likely. Given the area's intact nature, this infrastructure has a high probability of impacting bear use there. Collecting data on habitat use in the watersheds where this development will occur will be important to lessen impacts on critical use areas for bears. The higher elevations in this area are valuable denning habitat, so combining denning surveys and existing bear denning data to develop models identifying high-probability denning habitat can help direct development away from these areas.

As we move into the future, it will be important to understand how climate change will affect bears in the park. As the potentially warmer, wetter environment takes shape, the likelihood of changes in food availability, abundance, and available habitat could impact bears in the park, with both beneficial and detrimental impacts possible. Receding glaciers could allow for new habitat for bears at higher elevations. While new habitat is revealed, established vegetation communities will likely respond to changes in growing conditions, expanding and contracting to change the current landscape mosaic. Advancement of shrubs into alpine areas could supplant herbaceous plants and berries, diminishing the availability of these food sources. This change in community structure could be offset by increased precipitation that enhances plant productivity. Warmer conditions without increases in precipitation will cause plants to green and senesce early, affecting their availability to bears. Were this to occur in salt-marsh communities, the impacts on bears could be significant. Warming waters will have large effects on Alaska's cold-water fish, especially salmon. Unusually warm water temperatures during spawning runs could impede migratory corridors, cause pre-spawn mortality, and decrease reproductive success. In addition, warmer water temperature does influence developmental stages in salmon. They rear in freshwater, so warm water can reduce juvenile salmon growth rates, affecting survival due to freshwater and oceanic size-selective mortality. Acidification of the oceans is also occurring and will affect salmon, clams, mussels, and other marine species. Many of these are important foods for bears, so this development has the potential to greatly impact food availability in some systems.

NOTES

1. Cited in Branson, *Lake Clark–Iliamna, Alaska*, 69.
2. Branson, *Lake Clark–Iliamna, Alaska*, 75.
3. National Park Service, "Visitor Use."
4. National Park Service, "Right-of-Way Certificate of Access for North Tract of Johnson Tract."
5. Becker, "Brown Bear Line Transect Technique Development."
6. Schmidt et al., "Improving Inference for Aerial Surveys for Bears."
7. Schmidt et al., "Integrating Distance Sampling Survey Data with Population Indices."
8. Lafferty, Belant, and Phillips, "Testing the Niche Variation Hypothesis."
9. Robbins et al., "Optimizing Protein Intake as a Foraging Strategy."
10. Welch et al., "Constraints on Frugivory by Bears."
11. Robbins et al., "Optimizing Protein Intake as a Foraging Strategy."
12. Mangipane et al., "Dietary Plasticity in a Nutrient-Rich System."
13. Charnov, "Optimal Foraging"; Pyke, Pulliam, and Charnov, "Optimal Foraging."
14. Mangipane et al., "Influences of Landscape Heterogeneity on Home-Range Sizes of Brown Bears."
15. Libal et al., "Despotism and Risk of Infanticide."
16. Mangipane et al., "Sex-Specific Variation in Denning by Brown Bears."
17. Libal et al., "Despotism and Risk of Infanticide."
18. Libal et al., "Despotism and Risk of Infanticide."
19. Mangipane et al., "Sex-Specific Variation in Denning by Brown Bears."
20. National Park Service, "Visitor Use."
21. Nevin and Gilbert, "Perceived Risk, Displacement, and Refuging in Brown Bears"; Rode, Robbins, and Farley, "Sexual Dimorphism, Reproductive Strategy, and Human Activities."
22. Rode, Robbins, and Farley, "Behavioral Response of Brown Bears."

▶ Bear conservation in the future will need to factor in ecosystem changes due to a warming climate and the potential of coastal development. (Mark Johnson)

12

ECOLOGICAL VARIATION

RESILIENCE IN TIMES OF CHANGE

Grant V. Hilderbrand, Diana Lafferty, and Lindsey Mangipane

Throughout this book, we have been describing brown bears generally as a group, but a great deal of variation exists between individual bears and across bear populations. This "ecological plasticity" helps individual bears adjust to changing conditions, and the variation across individuals provides the substrate on which natural selection can act, enabling some individuals to persist or even thrive under changing conditions and thus buffering populations from the negative impacts of environmental change.

While research conducted in a specific park is critical to successful bear conservation and management decisions there, wider knowledge of brown bears across Alaska can tell us more about bear populations in different environments. For example, by comparing and contrasting the results of bear studies across multiple parks, we can improve our knowledge and better understand the level of ecological plasticity that exists in the species. Ecological plasticity is the ability to adapt to changing ecological conditions and be resilient to change. We conducted studies simultaneously in 3 national parks (Lake Clark, Katmai, and Gates of the Arctic) and Kodiak National Wildlife Refuge from 2014 through 2018. Because our study design was

◀ A young bear in Gates of the Arctic National Park and Preserve watches researchers. (NPS/Kyle Joly)

A great deal of variation exists among individual bears and across bear populations. Sizes, coloring, diets, and movement patterns show diverse patterns across Alaska. This variation, or ecological plasticity, may help them adapt to changing conditions in the future. (NPS/Nina Chambers)

replicated across these areas, we could compare our results, which enhanced our understanding of brown bears across diverse ecosystems by providing novel insights into brown bear ecology. It has been said that the best days of a scientist's career are when they are surprised by their findings. During these recent studies, we had many "best days."

IT IS MORE COMPLICATED THAN BROWN BEAR OR GRIZZLY BEAR

When considering brown bears at a continental scale, populations are often distinguished as those that rely heavily on salmon and those that don't.[1] Non-salmon-reliant bears are commonly called grizzlies, and coastal bears are referred to as brown bears. Bears from salmon-eating populations tend to be much larger, reproduce at a younger age, and have larger litters—all of which produce high population densities compared to bears from populations that have little or no access to salmon.[2] When we compared our 4 study areas, these trends certainly held true. For example, the muscle mass of the average female grizzly bear in Gates of the Arctic is around 175 pounds (80 kg). Brown bears in Lake Clark weigh in at around 235 pounds (105 kg) and at about 325 pounds (150 kg) in both Katmai and Kodiak.[3] Thus, the average muscle mass of bears in resource-rich coastal areas is almost twice as high as that of bears in the interior north. The average muscle mass of bears in Lake Clark, home to substantial Bristol Bay salmon runs, falls in between. The same patterns hold for other measurements used to evaluate structural body size, such as skull size and body length, for both male and female bears.[4] However, what was most striking to us was that individual variation *within* a given park was even more extreme than the variation seen *across* the different parks. For instance, muscle weight varied twofold within each park for females and threefold for males.[5]

In addition to variation in body size within populations, we also observed striking differences in the proportion of salmon in the diets of bears in Kodiak, Lake Clark, and Katmai.[6] Most surprising, some brown bears in Gates of the Arctic, which lies far inland and north of the Arctic Circle, consumed substantial amounts of salmon that contributed to more than half of their diet within a given year, while meat sources like moose and caribou contributed less than 10% to their late summer diet.[7] The detection of salmon in bear diets helped us document an additional 400 miles (650 km) of previously unknown anadromous streams in and around Gates of the Arctic.[8]

Another way we assess bears' resource needs is by evaluating their home range. Resource-rich coastal areas provide ample food for brown bears within a relatively small area, but interior grizzly bears have to travel farther to access the same amount of food. In our studies, we found that the home ranges of bears varied from a few square miles on the Katmai coast, a resource-rich environment,[9] to some of the largest home ranges ever quantified for bears in Gates of the Arctic, where resources are few and far between.[10] Lake Clark bear home ranges were more moderate in size, not as large as those observed in Gates of the Arctic and not as small as

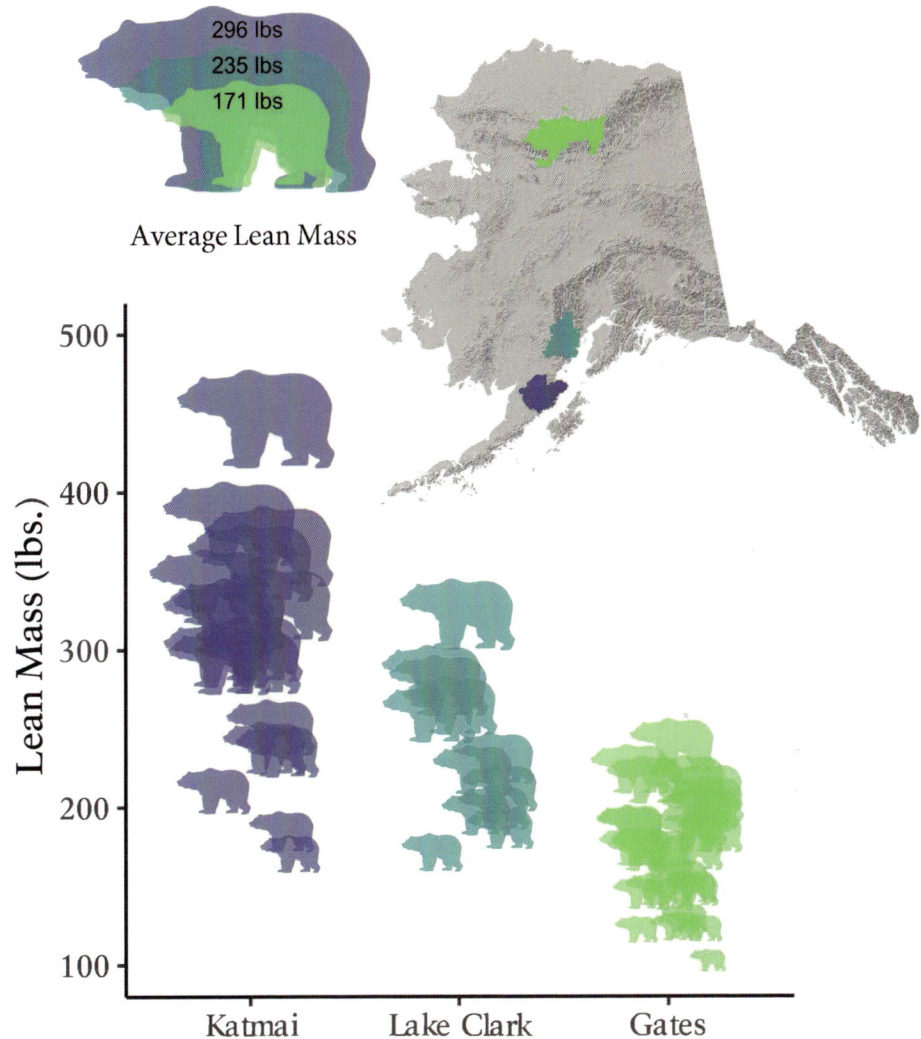

Average Lean Mass

296 lbs
235 lbs
171 lbs

Considerable variation exists in the spring body size of female brown bears both across and within national parks in Alaska. The values in this figure represent each bear's lean mass, which includes their muscle and skeleton but not their dramatically fluctuating body fat.

Bears are omnivores and have many strategies to get the calories and nutrition they need. Often, they eat a mix of meat and vegetation, capitalizing on food sources as they become locally abundant. (Troy Cambier, Chena River Aviation)

A FAT BEAR IS A HEALTHY BEAR, BUT THERE IS MORE THAN ONE WAY TO GET THERE

Across all wildlife species, survival and reproduction have costs, and the currency is calories. Prior work indicates that regardless of brown bear body size, a minimum threshold of about 20% body fat in the fall is necessary to support reproduction.[12] In addition, while some foods like salmon are incredibly rich in both fats and protein, brown bears actually benefit from consuming a combination of fat, protein, and carbohydrates.[13] Across our 4 study populations, bears' body condition in the fall tended to meet or surpass this threshold of 20% proportional body fat necessary to produce and support cubs regardless of diet or body size.[14] Although we observed a wide range of foraging strategies, variation in diet did not have a significant influence on important factors that contribute to an individual's ability to produce cubs, such as body condition or denning, perhaps because all the bears we sampled met their dietary needs.

A niche is generally defined at the position or function of an organism within its ecosystem. Brown bears across these different parks with variable diets provided an excellent opportunity to test the niche variation hypothesis (NVH).[15] The NVH is an ecological tenet that posits that greater variation across individuals in morphology (body

those observed in Katmai,[11] but both Gates of the Arctic and Lake Clark brown bear populations were characterized by diets that vary from one individual to the next with season and between males and females. We found that some bears were almost migratory, as they moved long distances from one seasonally available food resource to another. In other words, the population of bears in each park was not a collection of nearly identical-size animals roaming the landscape and using resources in the same way; rather, variation was the norm and exceptions were the rule.

shape and size) and diet offers individual bears an adaptive advantage.[16] Our findings generally support this hypothesis. What we observed was not just variation in bears from 1 population to the next but also wide variation across individuals within each population.[17] In essence, within each park, individual bears were making a living in a variety of ways, as reflected in their body size, diet, and habitat use. Niche variation resulted in an advantage directly to individuals but also to the population because individuals were not competing for the same resources in the same quantities. Rather, at the population level, bears functioned as generalist consumers, collectively using the full extent of resources available within the ecosystem, which maximized the population-level niche breadth. However, at the individual level, bears tended to be more selective in their resource use, exhibiting substantially narrower niche breadth compared to the population as a whole. Thus, the generalist bear populations within our study areas are actually composed of individuals with more specialized diets than what is represented by the population-level average. Individual specialization likely results in reduced competition among individuals, thereby enabling the population to be larger than what would be possible if all individuals were competing for all the same resources.

Niche Variation Hypothesis

Leigh Van Valen (August 12, 1935–October 16, 2010) was an evolutionary biologist in the Department of Ecology and Evolution at the University of Chicago, where he and his students conducted conceptually rich research to advance evolutionary theory. Though best known for his Red Queen Hypothesis, which posits the idea that a constant "arms race" is occurring between co-evolving species—a nod to the Red Queen's race in Lewis Carroll's 1871 novel *Through the Looking Glass*—over the past decade or so, Van Valen's niche variation hypothesis has had quite a research revival. In fact, NVH has inspired a new generation of scientists who are motivated not only to test fundamental ecological theory but also to apply NVH to understand better the diverse resource needs of species around the world.

Briefly, Van Valen theorized that populations occupying wider niches should exhibit greater among-individual variation compared to populations occupying narrower niches and that individual variation in dietary niche should confer an adaptive advantage.[18] As such, NVH provides a theoretical lens through which we can compare resource use within and between populations relative to measures of fitness like reproductive output or percentage body fat.

Importantly, as the combined effects of climate change and human disturbance continue to disrupt resource availability, NVH can provide a helpful framework for assessing whether changes in individual resource use impact population-level resource needs. This is important because wildlife are typically managed at a population level. However, if population-level averages do not adequately represent the diversity of resource use within the population, resource managers may not be able to adequately conserve the full range of resources used by the population.[19]

Coastal bears have access to marine food resources. Whale carcasses can provide a calorie boost for nearby bears. (NPS/Kelsey Griffin)

While there is an advantage to getting "big," as it allows for resource dominance and may minimize risk from attacks by other bears, there is also an increased cost to being large. In essence, if one has a big house, it costs more to heat it, cool it, and keep it running. Similarly, in times of limited resources (e.g., a berry crop failure, reduced salmon runs), larger bears may be more stressed as they struggle to meet their nutritional demands. Alternatively, smaller bears are more "efficient" and can build higher levels of fat than big bears while consuming the same amount of food. Therefore, rather than absolute body size driving the level of variation in the population, it appears that body condition (i.e., body fat) may be more important as individual females strive to meet the costs of cub production and rearing.

RESOURCE USE DYNAMICS BETWEEN BROWN BEARS AND BLACK BEARS

In addition to within-species interactions and variation in body size and diet, resource (or niche) overlap, competition, and exclusion can occur between black and brown bears. While brown bear body size in Denali is variable, similar to variability observed in other populations (e.g., Lake Clark), brown bear body condition and reproduction were not adversely affected by reduced salmon availability. Conversely, black bear body condition and reproduction *were* negatively impacted in low salmon years, suggesting that brown bears limit black bear access to salmon and thereby reduce the size of the niche (amount of resources) available to black bears when resources are scarce.[20] However, when we examined niche width in the same population of black and brown bears in this Denali study, we found that both species had similar body condition. This suggests to us that both species can successfully occupy a range of dietary niches while supporting the costs of survival and reproduction. These findings provide further support for the NVH across a dynamic ecosystem from a multi-species perspective.[21]

MICROBIAL HABITAT DIVERSITY

Stepping down from ecosystem-level processes and diversity, we have also assessed variation in the microbial communities in the digestive systems of bears from Katmai, Gates of the Arctic, and Lake Clark.[22] The digestive tract of individual bears is functionally a unique ecosystem that is shaped, in part, by the foods they eat. In addition, the community of microbiota in the gastrointestinal tract can influence a bear's ability to digest, absorb, and synthesize nutrients and respond to environmental stress; as such, it serves as an index of individual and population health. In general, gut microbial diversity was found to be greatest in Katmai and

Where brown bears and black bears overlap, they can select different dietary niches so they both receive the nutrition they need. These black bears in Glacier Bay National Park and Preserve are in a berry patch. The cubs have a locally specific coloring that enables them to better blend in with the rocky slopes carved by glaciers. (NPS/Cody Edwards)

lowest in Gates of the Arctic, likely tied to the diversity of foods available to and consumed by bears.[23] Beyond the diversity of microorganisms present in the digestive tract of bears, the composition of these microbial communities is important, too, as some organisms facilitate digestion and energy production whereas others may be opportunistic or even pathogenic.[24] However, because diet is a primary driver of gut microbial diversity in bears, changes in diet associated with global change may impact bear health. For example, consumption of processed human foods, which bears may access when garbage is not secured or if bait is available on the landscape to facilitate harvest, has been shown to impact gut microbial diversity.[25]

DIVERSITY ENHANCES FLEXIBILITY

As discussed in previous chapters, bear populations in each park face unique challenges associated with development in or adjacent to the park, park visitation, harvest, and other human impacts. In addition, the singular ecosystem-level (and global) perturbation facing bears, their habitat, and the resources on which they rely is climate change. Bears make a living by using the resources available to them on the landscape. As noted above, within a given park, not all bears use the same resources in the same quantities. By consuming different resources or different quantities of the same resources, bear populations are able to occupy all of the niches available to them collectively, and this supports more individual bears than would be possible if all bears consumed the same resources in the same quantities. As such, within-population variation in resource use increases the capacity of the population to withstand change, either long term (multi-year or decadal declines in salmon numbers) or short term (drought-caused berry crop failure). Further, some climate-related changes may be beneficial to bears, such as the expansion of the distribution or abundance of anadromous fish on the North Slope of Alaska and the lengthening of growing seasons; further, additional vegetation growth may increase the number of brown bears those areas can support. Alternatively, the effect of some climate-related changes may be very local and more difficult to predict, such as shifts in vegetation community composition that favor moose over caribou, thus changing the predominant protein sources available to bears. Simultaneously, increasing temperatures in freshwater streams have already been shown to adversely impact migrating salmon.[26]

The list of climate impacts on brown bears in Alaska parks is long, and the ecological affects are uncertain. That said, the physiological and behavioral flexibility observed in brown bears may aid their ability to respond to change both as individuals and as populations. However, plasticity has its limits, and we do not know if the changes in available niches, some lost and some gained, will result in an increased or decreased capacity of the land to support bears. Individual variation that allows for increased adaptability within a population is often observed at the ends of bell curves, yet most management focuses on the average animal in the population, in the middle of the curve. A failure to recognize the ecological value of specialist individuals by applying a one-size-fits-all management approach that focuses on conservation of the "average bear" may fail to conserve the resiliency of a population.[27]

In addition to designing research projects to address specific park management decisions, Alaska parklands can play a unique role in assessing the impacts of change, as they can serve as a "control" because their natural ecosystems are largely intact. Given the uncertainty about the magnitude of climate-related changes in the future, continued research and monitoring of bears and their habitat is crucial as the National Park Service continues to meet its mission of conserving the wildest of things in the wildest of places.

NOTES

1. Mowat and Heard, "Major Components of Grizzly Bear Diet across North America."
2. Hilderbrand et al., "Importance of Meat."
3. Hilderbrand et al., "Plasticity in Physiological Condition of Female Brown Bears."
4. Hilderbrand et al., "Body Size and Lean Mass of Brown Bears."
5. Hilderbrand et al., "Plasticity in Physiological Condition of Female Brown Bears"; Hilderbrand et al., "Body Size and Lean Mass of Brown Bears."
6. Deacy et al., "Kodiak Brown Bears Surf the Red Wave"; Mangipane et al., "Dietary Plasticity in a Nutrient-Rich System"; Erlenbach, "Nutritional and Landscape Ecology of Brown Bears."
7. Mangipane et al., "Dietary Plasticity and the Importance of Salmon."
8. Sorum et al., "Salmon Sleuths."
9. Erlenbach, "Nutritional and Landscape Ecology of Brown Bears."
10. Joly et al., "Factors Influencing Arctic Brown Bear Annual Home Range Sizes."
11. Mangipane et al., "Influences of Landscape Heterogeneity on Home-Range Sizes of Brown Bears."
12. Lopez-Alfaro et al., "Energetics of Hibernation and Reproductive Tradeoffs in Brown Bears."
13. Robbins et al., "Optimizing Protein Intake as a Foraging Strategy."
14. Hilderbrand et al., "Plasticity in Physiological Condition of Female Brown Bears"; Mangipane et al., "Dietary Plasticity in a Nutrient-Rich System."
15. Van Valen, "Morphological Variation and Width of Ecological Niche."
16. Van Valen, "Morphological Variation and Width of Ecological Niche."
17. Cameron et al., "Body Size Plasticity in North American Black and Brown Bears."
18. Van Valen, "Morphological Variation and Width of Ecological Niche."
19. Van Valen, "A New Evolutionary Law."
20. Belant et al., "Interspecific Resource Partitioning in Sympatric Ursids."
21. Lafferty, Belant, and Phillips, "Testing the Niche Variation Hypothesis."
22. Trujillo et al., "Intrinsic and Extrinsic Factors."

Continued research on bears and their habitat is crucial to conserving the wildest things in the wildest places. (NPS/Naomi Boak)

23. Trujillo et al., "Intrinsic and Extrinsic Factors."

24. Trujillo et al., "Correlating Gut Microbial Membership to Brown Bear Health Metrics."

25. Gillman, McKenney, and Lafferty, "Human-Provisioned Foods Reduce Gut Microbiome Diversity."

26. von Biela et al., "Evidence of Prevalent Heat Stress in Yukon River Chinook Salmon."

27. Darimont, Paquet, and Reimchen, "Landscape Heterogeneity and Marine Subsidy."

13

CONCLUSION

THE FUTURE OF BROWN BEAR CONSERVATION, MANAGEMENT, AND SCIENCE IN ALASKA'S PARKLANDS

Kyle Joly, David D. Gustine, Nina Chambers, and Grant V. Hilderbrand

Ever since the first humans crossed the Bering Land Bridge, they have shared Alaska with brown bears. While the relationship people and bears have forged has been fraught with difficulties, for millennia it has been marked by mutual respect and admiration. As we have shown throughout the book, this valuation of bears was shared among numerous Alaska Native groups across the state. While the "big animals'" sheer size and immense strength were certainly factors in gaining them respect, people have always also recognized the intelligence of brown bears.

This intelligence, combined with a healthy dose of curiosity, has helped brown bears occupy a wide array of habitats and niches—bears have found a lot of different ways to "make it work." While bears catching and eating spawning salmon as they leap up over Brooks Falls in Katmai National Park and Preserve may be the most common vision people think of, brown bears are generalists. They eat grasses, berries, insects, roots, ground squirrels, moose calves, carrion, clams, flounder, seals, and sea otters. Indeed, their nutritional requirements demand diverse diets, and bears have equally diverse strategies to address those

◄ Brown bears have been revered for thousands of years, not just for their impressive size and strength but also for their intelligence. Intelligence and curiosity have helped brown bears adapt to and thrive in diverse ecological conditions. (Stuart Leidner)

https://doi.org/10.5876/9781646427116.c013

needs: some bears live within a very small and constrained home range, while others range over seemingly boundless areas; some even appear to be migratory. Brown bears can be found from the temperate rainforests of southeast Alaska, near Klondike Gold Rush National Historical Park, north to the vast, treeless tundra of Gates of the Arctic National Park and Preserve.

And while this enormous variability and flexibility exist within the species, individual brown bears can also change their behaviors to adapt to the changing conditions they experience. For example, sows with cubs of the year may become more cautious and reduce their exposure to large boars that may try to kill their cubs. As a young boar matures, it may be confident enough to start fishing at prime locations. During a poor salmon run, individuals may switch to feeding more on berries and other vegetation or perhaps pursue human foods for bears that live near people. This heightened adaptability makes brown bears more resilient, yet it also amplifies the potential for humans and bears to interact and come into conflict.

Since 1918, the National Park Service (NPS) has managed brown habitat in Alaska. As early as 1931, the conservation of brown bear populations became an explicit goal of NPS management. In 1980, the Alaska National Interest Lands Conservation Act (ANILCA) vastly expanded the geographic scope and redoubled the NPS's focus on the conservation of this wilderness icon. The laws, regulations, and policies defining the mission of the NPS, along with its guiding philosophy, enable robust stewardship of brown bears. The conservation of large, intact ecosystems, where natural processes are allowed to play out, creates an ideal setting for brown bear populations.

A key pillar of conservation is striving to have enough knowledge about what you are trying to conserve so you can do it in a scientifically sound manner. The NPS, in conjunction with its many partners, has made significant scientific contributions that have expanded our collective knowledge of brown bears. We used this book as an opportunity to document as much NPS-related research and as many studies as we could so more people, like you, could learn about and appreciate the important roles brown bears have in Alaska's ecosystems and beyond.

What came as a surprise to us, the editors, was how much work had been done on brown bears on Alaska's national parklands. As biologists working with brown bears in parks, we thought we had a good grasp on the scope of the work that had been done. Why didn't we have a full understanding of the big picture? That in and of itself was one of the primary reasons we wrote this book.

Like many other wildlife management agencies, the NPS tends to compartmentalize as individual researchers and individual parks focus on their own efforts. Many biologists excel at completing field projects, but not all manage to publicize their work—not within the NPS or with the general public. Science done in a vacuum is much less impactful: knowledge needs to be shared to make the incremental improvements that are needed for conservation of wildlife.

In recent years, the coauthors of the chapters in this book have made great strides by collaborating with each other and across parks to undertake numerous comparative studies. We encourage further and broader collaborations in the future so we can continue to advance our understanding of brown bears in Alaska and throughout their global range. While there is much we know and have learned about brown bears through Alaska Native, local, and scientific knowledge, there is still much we don't know. Collaborations and comparative studies present excellent opportunities to continue to better our understanding and management of this species. Limited funding and rising costs of fieldwork and research will act as constraints on these efforts.

Modern humans present the greatest challenge for individual brown bears and the conservation of their species. These challenges are manifold, diverse, and often complex. Arguably, the biggest challenge is that of climate change, the effects of which are readily apparent across Alaska. How climate change will impact brown bears is not well-known and remains an important avenue for future research. Bears' innate ability to adjust to changing conditions and to use a wide array of habitats and resources should buffer them against some of the impacts of climate change. But climate change effects are happening so fast and are so intense that it is not entirely clear how bears will be affected. The role of warming ocean waters on salmon abundance, for example, could have dramatic effects on brown bear populations. Lengthening growing seasons and alterations to the timing and extent of snowfall and snowmelt will affect the denning period. Similarly, the impacts of wildfire, predicted to increase with warmer conditions, on brown bear habitat also remain poorly understood.

An ever-expanding human footprint on the landscape is another factor that influences brown bear populations and health. While bears are highly adaptable, they need large areas with relatively few people to thrive. Bears learn ways to avoid people where they can, but interactions between brown bears and humans often lead to negative outcomes for bears. These outcomes can include the loss or fragmentation of habitat, reduced access to primary foods, collisions with vehicles, harvest, and the removal of food-conditioned bears. Outside of parklands and even to some extent within them, development will continue as the human population continues to expand. Understanding the ecology and needs of brown bears and taking them

into account as development occurs will be vital for their conservation in the future.

Human visitation is another important management concern. Visitation to national parks, and specifically to see brown bears in Alaska, continues to increase. Viewing a wild brown bear in its natural environment within a park can be the most cherished moment of a trip to Alaska or even of a lifetime. The opportunity to have such experiences is critical to people valuing brown bears and, thus, to their conservation. Finding ways to balance visitors' experience and the needs of bears is complex. NPS law and policy dictates that when striving to balance between preserving resources and allowing people to enjoy them, preservation of the resource is the priority. This will be important for the NPS to keep in mind while managing for ever-increasing desire to see these magnificent creatures in their awe-inspiring natural habitats.

Changing human attitudes and perceptions of brown bears will also be a critical factor in the way brown bears fare in the coming decades. Reverence for "the big animal" has waned among segments of the population. Some people view brown bears as competition for ungulates (moose and caribou) and thus want to reduce their populations so they can have more hunting opportunities. Some fear them as dangerous creatures seeking out humans as prey—an over-hyped and very rare occurrence. Modern humans have the technological ability to heavily impact brown bears in numerous ways: our tolerance and restraint are required for their continued conservation. Conversely, increasing the public's understanding of brown bears will facilitate conservation efforts by affecting people's attitudes toward and perceptions of this fascinating animal.

This was the primary goal of this book: we hope you enjoyed learning more about Alaska brown bears and that you do your part to conserve them for the next seven generations . . . and beyond.

◄ Human attitudes toward brown bears are a critical factor in their conservation. Balancing visitation to view wild bears and keeping bear populations wild in the context of ever-expanding human impacts is a challenge. Our hope is that increasing public understanding and knowledge about bears will facilitate their conservation far into the future. (NPS/Lian Law)

BIBLIOGRAPHY

Ackerman, R. E. "Bluefish Cave." In F. H. West and C. F. West, eds., *American Beginnings: The Prehistory and Palaeoecology of Beringia*, 511–513. Chicago: University of Chicago Press, 1996.

Alaska Department of Fish and Game. *Anadromous Waters Catalog*. 2022. Accessed April 11, 2022. https://www.adfg.alaska.gov/sf/SARR/AWC/index.cfm?ADFG=main.interactive.

Alaska Department of Fish and Game. "Brown/Grizzly Bear Hunting in Alaska." 2021. Accessed April 11, 2022. http://www.adfg.alaska.gov/index.cfm%3Fadfg=brownbearhunting.main.

Alaska Department of Fish and Game. "Species Profile: Brown Bear." 2024. Accessed July 17, 2024. https://www.adfg.alaska.gov/index.cfm?adfg=brownbear.main.

Anderson, D. B., W. W. Anderson, R. Bane, R. K. Nelson, and N. S. Towarak. *Kuuvaŋmiut Subsistence: Traditional Eskimo Life in the Later Twentieth Century*. Kotzebue, AK: National Park Service, United States Department of the Interior, 1998.

Becker, E. F. "Brown Bear Line Transect Technique Development, 1 July 1999–30 June 2000." Federal Aid in Wildlife Restoration Research Performance Report, Grant W-27-3. Alaska Department of Fish and Game, Juneau, 2001.

Becker, E. F., and A. M. Christ. "A Unimodal Model for Double Observer Distance Sampling Surveys." *PLoS ONE* 10 (2015): e0136403.

Becker, E. F., and D. W. Crowley. "Estimating Brown Bear Abundance and Harvest Rate on the Southern Alaska Peninsula." *PLoS ONE* 16 (2021): e0245367.

Becker, E. F., and P. X. Quang. "A Gamma-Shaped Detection Function for Line-Transect Surveys with Mark-Recapture and Covariate Data." *Journal of Agricultural, Biological, and Environmental Statistics* 14 (2009): 207–223.

Behnke, S. R. *Subsistence Use of Brown Bear in the Bristol Bay Area: A Review of Available Information*. Dillingham: Alaska Department of Fish and Game, Division of Subsistence, 1981.

Belant, J. L. "Resource Partitioning by Sympatric Brown and American Black Bears." PhD diss., University of Alaska, Fairbanks, 2006.

Cubs tussle in the shallows. (NPS/Jim Pfeiffenberger)

Belant, J. L. *Resource Selection of Brown Bears and Black Bears in Southcentral Alaska: A Progress Report*. Denali National Park, AK: Denali National Park and Preserve, 1999.

Belant, J. L., K. Kielland, E. H. Follmann, and L. G. Adams. "Interspecific Resource Partitioning in Sympatric Ursids." *Ecological Applications* 16 (2006): 2333–2343.

Benn, B., and S. Herrero. "Grizzly Bear Mortality and Human Access in Banff and Yoho National Parks, 1971–98." *Ursus* 13 (2002): 213–221.

Birkedal, T. "Ancient Hunters in the Alaskan Wilderness: Human Predators and Their Role and Effect on Wildlife Populations and the Implications for Resource Management." Paper presented at the Cultural Perspectives on Wilderness session, Seventh George Wright Society Meeting, National Park Service Alaska Region, Jacksonville, FL, November 16–20, 1992.

Bled, F., J. L. Belant, L. J. Van Daele, N. Svoboda, D. D. Gustine, G. V. Hilderbrand, and V. G. Barnes. "Using Multiple Data Types and Integrated Population Models to Improve Our Knowledge of Apex Predator Population Dynamics." *Ecology and Evolution* (2017). https://doi.org/10.1002/ece3.3469.

Boulanger, J., and G. B. Stenhouse. "The Impact of Roads on the Demography of Grizzly Bears in Alberta." *PLoS ONE* (2017). https://doi.org/10.1371/journal.pone.0115535.

Bowen, L., A. K. Miles, S. C. Waters, D. D. Gustine, K. Joly, and G. V. Hilderbrand. "Using Gene Transcription to Assess Ecological and Anthropological Stressors in Brown Bears." *EcoHealth* 15, no. 1 (2017). https://doi.org/10.1007/s10393-017-1287-0.

Box, J. E., W. T. Colgan, T. R. Christensen, N. M. Schmidt, M. Lund, F. J. W. Parmentier, R. Brown, U. S. Bhatt, E. S. Euskirchen, V. E. Romanovsky et al. "Key Indicators of Arctic Climate Change: 1971–2017." *Environmental Research Letters* 14, no. 4 (2019): 045010.

Braaten, A. M., and B. K. Gilbert. *Profile Analysis of Human-Bear Relationships in Katmai National Park and Preserve*. Anchorage: National Park Service, 1986.

Branson, J., ed. *Lake Clark–Iliamna, Alaska, 1921: The Travel Diary of Colonel A. J. Macnab, with Related Documents*. Anchorage: Alaska Natural History Association, 1996.

Cameron, M. D., G. V. Hilderbrand, K. Joly, J. H. Schmidt, D. D. Gustine, L. S. Mangipane, B. A. Mangipane, and M. S. Sorum. "Body Size Plasticity in North American Black and Brown Bears." *Ecosphere* 11, no. 8 (2020): e03235. https://doi.org/10.1002/ecs2.3235.

Cathcart, C. N., and J. Giefer. "Fish Inventories of the Upper Kobuk and Koyukuk River Basins." *Alaska Park Science* 19, no. 1 (2020): 10–17.

Charnov, E. L. "Optimal Foraging: The Marginal Value Theorem." *Theoretical Population Biology* 9 (1976): 129–136.

Clemens, J., and F. Norris. *Building in an Ashen Land: Historic Resource Study of Katmai National Park and Preserve*. Anchorage: National Park Service, Alaska Support Office, 2008.

Coffing, M. W. *Subsistence Use of Brown Bear in Western Alaska*. Juneau: Alaska Department of Fish and Game, Division of Subsistence, 1992.

Dalle-Molle, J. L. *Field Tests and Users' Opinions of Bear Resistant Backpack Food Containers in Denali National Park, AK, 1982 and 1983*. Anchorage: National Park Service, 1984.

Dalle-Molle, J. L., and J. C. Van Horn. "Bear-People Conflict Management in Denali National Park." In *Proceedings of a Symposium on Management Strategies*, 121–127. Inuvik, AK: Northwest Territories Department of Renewable Resources, 1989.

Darimont, C. T., P. C. Paquet, and T. E. Reimchen. "Landscape Heterogeneity and Marine Subsidy Generate Extensive Intrapopulation Niche Diversity in a Large Terrestrial Vertebrate." *Journal of Animal Ecology* (2009): 126–133.

Dauenhauer, N., and R. Dauenhauer, eds. *Haa Shuká, Our Ancestors: Tlingit Oral Narratives*. Juneau, AK: Sealaska Heritage Foundation, 1987.

Deacy, W. W., J. B. Armstrong, W. B. Leacock, and J. A. Stanford. "Phenological Synchronization Disrupts Trophic Interactions between Kodiak Brown Bears and Salmon." *Proceedings of the National Academy of Sciences* 114 (2017): 10432–10437.

Deacy, W. W., W. B. Leacock, J. B. Armstrong, and J. A. Stanford. "Kodiak Brown Bears Surf the Red Wave: Direct Evidence from GPS Collared Individuals." *Ecology* (2016). https://doi.org/10.1890/15-1060.1.

Dean, F. "Brown Bear Density, Denali National Park, Alaska, and Sighting Efficiency Adjustment." *International Conference on Bear Research and Management* 7 (1987): 37–43.

DeBruyn, T. D., T. S. Smith, K. Proffitt, S. Partridge, and T. D. Drummer. "Brown Bear Response to Elevated Viewing Structures at Brooks River, Alaska." *Wildlife Society Bulletin* 32 (2004): 1132–1140.

Dekinaak. "Origin of Iceberg House." In J.R. Swanton, ed., *Tlingit Myths and Texts*, 52–53. Washington, Government Printing Office, 1909.

de Laguna, F. "The Atna of the Copper River, Alaska: The World of Men and Animals." *Folk* 11–12 (1969): 17–26.

de Laguna, F. *Under Mount Saint Elias: The History and Culture of the Yakutat Tlingit*. Smithsonian Contributions to Anthropology 7, 1–3. Washington, DC: Smithsonian Institution, 1972.

DeVoto, B., ed. *The Journals of Lewis and Clark*. Boston: Houghton Mifflin, 1953.

Ditmer, M. A., J. B. Vincent, L. K. Werden, J. C. Tanner, T. G. Laske, P. A. Iaizzo, D. L. Garshelis, and J. R. Fieberg. "Bears Show a Physiological but Limited Behavioral Response to Unmanned Aerial Vehicles." *Current Biology* 25 (2015): 2278–2283.

Dixon, J. S. *Birds and Mammals of Mount McKinley National Park, AK*. National Park Service Fauna Series 3. Washington, DC: Government Printing Office, 1938.

Dumond, D. E. 2011. *Archaeology on the Alaska Peninsula: The Northern Section, Fifty Years Onward*. Anthropological Papers 70. Eugene: University of Oregon, 2011.

Emmons, G. T. *The Tlingit Indians*. Edited with additions by Frederica de Laguna. American Museum of Natural History, Anthropological Paper 70. Seattle: University of Washington Press and the American Museum of Natural History, 1991.

Erlenbach, J. A. "Nutritional and Landscape Ecology of Brown Bears in Katmai National Park, Alaska." PhD diss., Washington State University, Pullman, 2020.

Erlenbach, J. A., K. R. Griffin, and C. T. Robbins. *The Need for Habitat-Specific Management of Bears and Bear Viewing*. Final report provided to the National Park Service, Anchorage, 2021.

Fall, J. A., and L. B. Hutchinson-Scarbrough. *Subsistence Uses of Brown Bears in Communities of Game Management Unit 9E, Alaska Peninsula, Southwest Alaska*. Juneau: Alaska Department of Fish and Game, Division of Subsistence, 1996.

Fortin, J. K., K. D. Rode, G. V. Hilderbrand, J. Wilder, S. Farley, C. Jorgensen, and B. G. Marcot. "Impacts of Human Recreation on Brown Bears (*Ursus arctos*): A Review and New Management Tool." *PLoS ONE* 11 (2016): e0141983.

French, H. B. "Effects of Bear Viewers and Photographers on Brown Bears (*Ursus arctos*) at Hallo Bay, Katmai National Park and Preserve, Alaska." MS thesis, University of Alaska Fairbanks, 2007.

Georgette, S. *Brown Bears on the Northern Seward Peninsula, Alaska: Traditional Knowledge and Subsistence Uses*. Technical Paper 248. Juneau: Alaska Department of Fish and Game, Division of Subsistence, 2001.

Gerlach, C., M. Newman, E. J. Knell, and E. S. Hall Jr. "Blood Protein Residues on Lithic Artifacts from Two Archaeological Sites in the De Long Mountains, Northwestern Alaska." *Arctic* 49, no. 1 (1996): 1–10.

Germonpre, M., and R. Hamalainen. "Fossil Bear Bones in the Belgian Upper Paleolithic: The Possibility of a Proto Bear-Ceremonialism." *Arctic Anthropology* 44, no. 2 (2007): 1–30.

Gillman, S. J., E. A. McKenney, and D. J. R. Lafferty. "Human-Provisioned Foods Reduce Gut Microbiome Diversity in American Black Bears (*Ursus americanus*)." *Journal of Mammalogy* 103, no. 2 (2021): 339–346. https://doi.org/10.1093/jmammal/gyab154.

Glitzenstein, E., and J. Fritschie. "The Forest Service's Bait and Switch: A Case Study on Bear Baiting and the Service's Struggle to Adopt a Reasoned Policy on a Controversial Hunting Practice within the National Forests." *Animal Law* 47, no. 56 (1995): 47–81.

Glover, J. M. *A Wilderness Original*. Seattle: The Mountaineers, 1986.

González-Bernardo, E., L. F. Russo, E. Valderrábano, Á. Fernández, and V. Penteriani. *Denning in Brown Bears*. Hoboken, NJ: John Wiley and Sons, 2020.

Griffin, K. R. *Spatio-Temporal Distribution of Coastal Brown Bears and Visitors in Katmai National Park, Alaska*. Natural Resource Report NPS/KATM/NRR—2021/2216. Fort Collins, CO: National Park Service, 2021. https://doi.org/10.36967/nrr-2283864.

Gunther, K. A., and M. A. Haroldson. "Potential for Recreational Restrictions to Reduce Grizzly Bear–Caused Human Injuries." *Ursus* 31 (2020): 1–17.

Haroldson, M. A., and K. A. Gunther. "Roadside Bear Viewing Opportunities in Yellowstone National Park: Characteristics, Trends, and Influence of Whitebark Pine." *Ursus* 24 (2013): 27–41.

Haugen, M. B. "Blood Concentrations of Lead (Pb), Mercury (Hg), and Cadmium (Cd) in Scandinavian and Alaskan Brown Bears (*Ursus arctos*)." Master's thesis, Inland Norway University, Rena, 2020.

Herrero, S. *Bear Attacks: Their Causes and Avoidance*. Guilford, CT: Lyons, 2018.

Herrero, S., T. Smith, T. D. DeBruyn, K. Gunther, and C. A. Matt. "From the Field: Brown Bear Habituation to People—Safety, Risks, and Benefits." *Wildlife Society Bulletin* 33 (2005): 1–12.

Hilderbrand, G. V., D. D. Gustine, K. Joly, B. A. Mangipane, W. Leacock, M. D. Cameron, M. S. Sorum, L. S. Mangipane, and J. A. Erlenbach. "Influence of Maternal Body Size, Condition, and Age on Recruitment of Four Alaska Brown Bear Populations." *Ursus* 29, no. 2 (2019): 111–118. https://doi.org/10.2192/URSUS-D-18-00008.1.

Hilderbrand, G. V., D. D. Gustine, B. A. Mangipane, K. Joly, W. Leacock, L. S. Mangipane, J. A. Erlenbach, M. S. Sorum, M. D. Cameron, J. L. Belant, and T. Cambier. "Body Size and Lean Mass of Brown Bears across and within Four Diverse Ecosystems." *Journal of Zoology* 305 (2018): 53–62. https://doi.org/10.1111/jzo.12536.

Hilderbrand, G. V., D. D. Gustine, B. A. Mangipane, K. Joly, W. Leacock, L. S. Mangipane, J. A. Erlenbach, M. S. Sorum, M. D. Cameron, J. L. Belant, and T. Cambier. "Plasticity in Physiological Condition of Female Brown Bears across Diverse Ecosystems." *Polar Biology* 41, no. 4 (2018): 773–780. https://doi.org/10.1007/s00300-017-2238-5.

Hilderbrand, G. V., T. A. Hanley, C. T. Robbins, and C. C. Schwartz. "Role of Brown Bears (*Ursus arctos*) in the Flow of Marine Nitrogen in a Terrestrial Ecosystem." *Oecologia* 121 (1999): 546–550.

Hilderbrand, G. V., S. G. Jenkins, C. C. Schwartz, T. A. Hanley, and C. T. Robbins. "Effect of Seasonal Differences in Dietary Meat Intake on Changes in Body Mass and Composition in Wild and Captive Brown Bears." *Canadian Journal of Zoology* 77 (1999): 1623–1630.

Hilderbrand, G. V., K. Joly, M. S. Sorum, M. D. Cameron, and D. D. Gustine. "Brown Bear (*Ursus arctos*) Body Size, Condition, and Productivity in the Arctic, 1977–2016." *Polar Biology* 42 (2019): 1122–1130. https://doi.org/10.1007/s00300-019-02501-8.

Hilderbrand, G. V., C. C. Schwartz, C. T. Robbins, M. E. Jacoby, T. A. Hanley, S. M. Arthur, and C. Servheen. "The Importance of Meat, Particularly Salmon, to Body Size, Population Productivity, and Conservation of North American Brown Bears." *Canadian Journal of Zoology* 77 (1999): 132–138.

Hopkins, J. B., S. Herrero, R. T. Shideler, K. A. Gunther, C. C. Schwartz, and S. T. Kalinowski. "A Proposed Lexicon of Terms and Concepts for Human-Bear Management in North America." *Ursus* 21 (2010): 154–168.

Howe, J. R. *Bear Man of Admiralty Island: A Biography of Allen E. Hasselborg*. Fairbanks: University of Alaska Press, 1996.

Intergovernmental Panel on Climate Change (IPCC). "Climate Change 2021 The Physical Science Basis Summary for Policymakers." *Climate Change 2021: The Physical Science Basis* (2021): 3949.

Jacobs, M., Jr., and M. Jacobs Sr. "Southeast Alaska Native Foods." In A. Hope III, ed., *Raven's Bones*, 112–130. Sitka, AK: Sitka Community Association, 1982.

Joly, K., M. D. Cameron, M. S. Sorum, D. D. Gustine, W. Deacy, and G. V. Hilderbrand. "Factors Influencing Arctic Brown Bear Annual Home Range Sizes and Limitations of Home Range Analyses." *Ursus* 33 (2022): e11. https://doi.org/10.2192/URSUS-D-21-00015.2.

Joly, K., P. A. Duffy, and T. S. Rupp. "Simulating the Effects of Climate Change on Fire Regimes in Arctic Biomes: Implications for Caribou and Moose Habitat." *Ecosphere* 3 (2012): art36.

Kamenskii, Fr. Anatoli. *Tlingit Indians of Alaska*. Translated by Sergei Kan. Rasmuson Library Historical Translation Series, vol. 2. Fairbanks: University of Alaska Press, 1985 [1906].

Kasank, *The Man Who Entertained the Bears*, 220–222.

Kasischke, E. S., D. L. Verbyla, T. S. Rupp, A. D. McGuire, K. A. Murphy, R. Jandt, J. L. Barnes, E. E. Hoy, P. A. Duffy, M. Calef, and M. R. Turetsky. "Alaska's Changing Fire Regime: Implications for the Vulnerability of Its Boreal Forests." *Canadian Journal of Forestry Research* 40 (2010): 1313–1324.

Kasworm, W. F., and T. L. Manley. "Road and Trail Influences on Grizzly Bears and Black Bears in Northwest Montana." *International Conference on Bear Research and Management* 8 (1990): 79–84.

Keay, J. A. *Grizzly Bear Population Ecology and Monitoring, Denali National Park and Preserve, Alaska: Report of Project Development and Findings, 2J001*. Anchorage: Alaska Biological Science Center, 2001.

Kendall, K. C., T. A. Graves, J. A. Royle, A. C. Macleod, K. S. McKelvey, J. Boulanger, and J. S. Waller. "Using Bear Rub Data and Spatial Capture-Recapture Models to Estimate Trends in a Brown Bear Population." *Scientific Reports* 9, no. 1 (2019): 16804.

Kendall, K. C., J. B. Stetz, D. A. Roon, L. P. Waits, J. B. Boulanger, and D. Paetkau. "Grizzly Bear Density in Glacier National Park, Montana." *Journal of Wildlife Management* 72 (2008): 1693–1705.

Kendall, S. Interview. September 7, 2017. Glacier Bay National Park and Preserve Collection, Gustavus, AK.

Kopperl, R. E. "Cultural Complexity and Resource Intensification on Kodiak Island, Alaska." PhD diss., University of Washington, Seattle, 2003.

Krauss, M. E., G. Holton, J. Kerr, and C. West. *Indigenous Peoples and Languages of Alaska*. Fairbanks: Alaska Native Language Center and UAA Institute of Social and Economic Research, 2011. Online: https://www.uaf.edu/anla/collections/map/.

Lackey, C., D. Telesco, K. Annis, D. Battle, H. Cooley, P. Frame, L. Mangipane, C. Olfenbuttel, M. Vieira, and T. Waldrop. In press. *Ursus*.

Lafferty, D. J., J. L. Belant, and D. L. Phillips. "Testing the Niche Variation Hypothesis with a Measure of Body Condition." *Oikos* 124 (2015): 732–740.

Lamb, C. T., G. Mowat, A. Reid, L. Smit, M. Proctor, B. N. McLellan, S. E. Nielsen, and S. Boutin. "Effects of Habitat Quality and Access Management on the Density of a Recovering Grizzly Bear Population." *Journal of Applied Ecology* 55 (2017): 1406–1417.

Levi, T., G. Hilderbrand, M. Hocking, T. Quinn, K. White, M. Adams, J. Armstrong, A. Crupi, C. Darimont, W. Deacy et al. "Community Ecology and Conservation of Bear-Salmon Ecosystems." *Frontiers in Ecology and Evolution* 8 (2020): 1–16.

Lewis, T. M. "Shoreline Distribution and Landscape Genetics of Bears in a Recently Deglaciated Fjord: Glacier Bay, Alaska." MS thesis, University of Alaska Fairbanks, Fairbanks, 2021.

Lewis, T. M., and D. J. R. Lafferty. "Brown Bears and Wolves Scavenge Humpback Whale Carcass in Alaska." *Ursus* 25 (2014): 8–13. https://doi.org/10.2192/URSUS-D-14-00004.1.

Lewis, T. M., S. Pyare, and K. J. Hundertmark. "Contemporary Genetic Structure of Brown Bears (*Ursus arctos*) in a Recently Deglaciated Landscape." *Journal of Biogeography* 42 (2015): 1701–1713. https://doi.org/10.1111/jbi.12524.

Lewis, T. M., A. E. Stanek, and K. B. Young. *Bears in Glacier Bay National Park and Preserve: Sightings, Human Interactions, and Research 2010–2017*. Natural Resource Report NPS/GLBA/NRR—2020/2134. Fort Collins, CO: National Park Service, 2020.

Libal, N. S., J. L. Belant, B. D. Leopold, G. Wang, and P. A. Owen. "Despotism and Risk of Infanticide Influence Grizzly Bear Den-Site Selection." *PLoS ONE* 6 (2011): e24133.

Loon, H., and S. Georgette. *Contemporary Brown Bear Use in Northwest Alaska*. Technical Paper 163. Kotzebue: Alaska Department of Fish and Game, Division of Subsistence, 1989.

Lopez-Alfaro, C., C. T. Robbins, A. Zedrosser, and S. E. Nielsen. "Energetics of Hibernation and Reproductive Tradeoffs in Brown Bears." *Ecological Modelling* 270 (2013): 1–10.

Loso, M. G., C. F. Larsen, B. S. Tober, M. Christoffersen, M. Fahnestock, J. W. Holt, and M. Truffer. "Quo Vadis, Alsek? Climate-Driven Glacier Retreat May Change the Course of a Major River Outlet in Southern Alaska." *Geomorphology* 384 (2021): 107701.

Loveless, K., T. Olson, T. Hamon, and L. Butler. "Population Assessment of Brown Bears in Katmai National Preserve, Alaska." Unpublished document, 2009.

Luehrmann, S. *Alutiiq Villages under Russian and U.S. Rule*. Fairbanks: University of Alaska Press, 2009.

Mace, R. D., L. Phillips, T. Meier, and P. Owen. *Habitat Use and Movement Patterns of Grizzly Bears in Denali National Park Relative to the Denali Park Road*. Natural Resource Technical Report NPS/DENA/NRTR—2012/563. Fort Collins, CO: National Park Service, 2012.

Mace, R. D., J. S. Waller, T. L. Manley, L. J. Lyon, and H. Zuuring. "Relationships among Grizzly Bears, Roads, and Habitat in the Swan Mountains, Montana." *Journal of Applied Ecology* 33, no. 6 (1996): 1395–1404. https://www.jstor.org/stable/2404779.

Malick, M. J., and S. P. Cox. "Regional-Scale Declines in Productivity of Pink and Chum Salmon Stocks in Western North America." *PLoS ONE* 11 (2016): e0146009.

Mangipane, L. S., J. L. Belant, T. L. Hiller, M. E. Colvin, D. D. Gustine, B. A. Mangipane, and G. V. Hilderbrand. "Influences of Landscape Heterogeneity on Home-Range Sizes of Brown Bears." *Mammalian Biology* 88 (2017): 1–7. https://doi.org/10.1016/j.mambio.2017.09.002.

Mangipane, L. S., J. L. Belant, D. J. R. Lafferty, D. D. Gustine, T. L. Hiller, M. E. Colvin, B. A. Mangipane, and G. V. Hilderbrand. "Dietary Plasticity in a Nutrient-Rich System Does Not Influence Brown Bear (*Ursus arctos*) Body Condition or Denning." *Polar Biology* 41 (2018): 763–772.

Mangipane, L. S., J. L. Belant, B. A. Mangipane, D. D. Gustine, and G. V. Hilderbrand. "Sex-Specific Variation in Denning by Brown Bears." *Mammalian Biology* 93 (2018): 38–44.

Mangipane, L. S., D. J. R. Lafferty, K. Joly, M. S. Sorum, M. D. Cameron, J. L. Belant, G. V. Hilderbrand, and D. D. Gustine. "Dietary

Plasticity and the Importance of Salmon to Brown Bear (*Ursus arctos*) Body Size and Condition in a Low Arctic Ecosystem." *Polar Biology* 43 (2020): 825–833. https://doi.org/10.1007/s00300-020-02690-7.

Marcot, B. G., M. T. Jorgenson, J. P. Lawler, C. M. Handel, and A. R. DeGange. "Projected Changes in Wildlife Habitats in Arctic Natural Areas of Northwest Alaska." *Climatic Change* 130 (2015): 145–154.

Mattson, D. J., S. Herrero, and T. Merrill. "Are Black Bears a Factor in the Restoration of North American Grizzly Bear Populations?" *Ursus* 16 (2005): 11–30. https://doi.org/10.1016/0041–3879(74)90069–5.

Mattson, D. J., R. R. Knight, and B. M. Blanchard. "The Effects of Developments and Primary Roads on Grizzly Bear Habitat Use in Yellowstone National Park, Wyoming." *International Conference on Bear Research and Management* 7 (1987): 259–273.

McLaren, D., R. J. Wigen, Q. Mackie, and D. W. Fedje. "Bear Hunting at the Pleistocene/Holocene Transition on the Northern Northwest Coast of North America." *Canadian Zooarchaeology* 22 (2005): 3–29.

McLellan, B. N., and D. M. Shackleton. "Grizzly Bears and Resource Extraction Industries: Effects of Roads on Behavior, Habitat Use, and Demography." *Journal of Applied Ecology* 25 (1988): 451–460.

Miller, S. D., G. C. White, R. A. Sellers, H. V. Reynolds, J. W. Schoen, K. Titus, V. G. Barnes Jr., R. B. Smith, R. R. Nelson, W. B. Ballard, and C. C. Schwartz. "Brown and Black Bear Density Estimation in Alaska Using Radiotelemetry and Replicated Mark-Resight Techniques." *Wildlife Monographs* 133 (1997): 3–55.

Milner, A. M., E. E. Knudsen, C. Soiseth, A. L. Robertson, D. Schell, I. T. Phillips, and K. Magnusson. "Colonization and Development of Stream Communities across a 200-Year Gradient in Glacier Bay National Park, Alaska, USA." *Canadian Journal of Fisheries and Aquatic Sciences* 57 (2000): 2319–2335.

Monson, D. H., R. L. Taylor, G. V. Hilderbrand, J. A. Erlenbach, H. A. Coletti, K. A. Kloecker, G. G. Esslinger, and J. L. Bodkin. "Brown Bear–Sea Otter Interactions along the Katmai Coast." *Journal of Mammalogy* 104, no. 1 (2023): 171–183. https://doi.org/10.1093/jmammal/gyac095.

Morton, J. M., G. C. White, G. D. Hayward, D. Paetkau, and M. P. Bray. "Estimation of the Brown Bear Population on the Kenai Peninsula, Alaska." *Journal of Wildlife Management* 80 (2016): 332–346.

Mowat, G., and D. C. Heard. "Major Components of Grizzly Bear Diet across North America." *Canadian Journal of Zoology* 84 (2011): 473–489.

Murie, A. *The Grizzlies of Mount McKinley*. Scientific Monograph Series 14. Washington, DC: United States Department of the Interior, National Park Service, 1981.

National Park Service. *Bear Safety in Alaska's National Parklands*. N.d. Accessed April 11, 2022. https://www.nps.gov/wrst/planyourvisit/upload/bear-safe-brochure.pdf.

National Park Service. *Glacier Bay Bear-Human Management Plan, Bartlett Cove, Alaska*. Anchorage: National Park Service, 2013.

National Park Service. "Hunting and Trapping in National Preserves: Final Rule." *Federal Register* 85 (2020): 35181–35191.

National Park Service. *Management Policies 2006*, chapter 4, "Natural Resource Management." Washington, DC: United States Department of the Interior, 2006.

National Park Service. "Right-of-Way Certificate of Access for North Tract of Johnson Tract, Environmental Assessment." United States Department of the Interior, Lake Clark National Park and Preserve, Alaska Region, September 2020.

National Park Service. "Visitor Use." 2021. Accessed April 11, 2022. https://www.nps.gov/articles/visitor-use.htm.

National Park Service. "Visitor Use, Katmai National Park and Preserve, Lake Clark National Park and Preserve." 2021. Accessed July 17, 2024. https://www.nps.gov/articles/visitor-use.htm.

National Park Service and Alaska Department of Fish and Game. *Best Practices for Viewing Bears on the West Side of Cook Inlet and the Katmai Coast*. King Salmon, AK: National Park Service, 2003.

Nelson, R. K. *Make Prayers to the Raven: A Koyukon View of the Northern Forest*. Chicago: University of Chicago Press, 1983.

Nelson, R. K., K. H. Mautner, and G. R. Bane. *Tracks in the Wildland: A Portrayal of Koyukon and Nunamiut Subsistence*. Fairbanks: University of Alaska, Anthropology and Historic Preservation, Cooperative Park Studies Unit, 1982.

Nettles, J., M. Brownlee, R. Sharp, S. Jackson, and D. Dagan. *Evaluation of the Bear Viewing Experience and Associated Thresholds at Katmai National Park and Preserve and Lake Clark National Park and Preserve, 2017–2020*. Research Report, Clemson University, Clemson, SC, 2020.

Nevin, O. T., and B. K. Gilbert. "Perceived Risk, Displacement, and Refuging in Brown Bears: Positive Impacts of Ecotourism?" *Biological Conservation* 121 (2005): 611–622.

Newton, R., and M. Moss. *The Subsistence Lifeway of Tlingit People: Excerpts of Oral Interviews.* Tongass National Forest, Document 131. Juneau: USDA Forest Service, 1987.

Olson, R. L. *The Social Structure and Social Life of the Tlingit in Alaska.* Anthropological Records 26. Berkeley: University of California, 1967.

Olson, T. L., and B. K. Gilbert. "Variable Impacts of People on Brown Bear Use of an Alaskan River." *Bears: Their Biology and Management* 9 (1994): 97–106.

Olson, T. L., B. K. Gilbert, and R. C. Squibb. "The Effects of Increasing Human Activity on Brown Bear Use of an Alaskan River." *Biological Conservation* 82 (1997): 95–99.

Olson, T. L., and J. A. Putera. *Refining Monitoring Protocols to Survey Brown Bear Populations in Katmai National Park and Preserve and Lake Clark National Park and Preserve.* Anchorage: National Park Service, 2007.

Olson, T. L., R. C. Squibb, and B. K. Gilbert. "Brown Bear Diurnal Activity and Human Use: A Comparison of Two Salmon Streams." *Ursus* 10 (1998): 547–555.

Parlee, B., and K. J. Caine. *When the Caribou Do Not Come: Indigenous Knowledge and Adaptive Management in the Western Arctic.* Vancouver: University of British Columbia Press, 2018.

Partlow, M. *Salmon Intensification and Changing Household Organization in the Kodiak Archipelago.* PhD diss., University of Wisconsin–Madison, Bell and Howell Information and Learning Company, Ann Arbor, MI, 2000.

Partridge, S., T. Smith, and T. M. Lewis. *Black and Brown Bear Activity at Selected Coastal Sites in Glacier Bay National Park and Preserve, Alaska: A Preliminary Assessment using Noninvasive Procedures.* Open-File Report 2009–1169. Anchorage: United States Geological Survey, 2009.

Penteriani, V., J. V. López-Bao, C. Bettega, F. Dalerum, M. del Mar Delgado, K. Jerina, I. Kojola, M. Krofel, and A. Ordiz. "Consequences of Brown Bear Viewing Tourism: A Review." *Biological Conservation* 206 (2017): 169–180.

Piatt, J. F., J. K. Parrish, H. M. Renner, S. K. Schoen, T. T. Jones, M. L. Arimitsu, K. J. Kuletz, B. Bodenstein, M. García-Reyes, R. S. Duerr et al. "Extreme Mortality and Reproductive Failure of Common Murres Resulting from the Northeast Pacific Marine Heatwave of 2014–2016." *PLoS ONE* 15, no. 1 (2020): e0226087. https://doi.org/10.1371/journal.pone.0226087.

Pinjuv, K. "Estimating Black Bear Population Size in Gustavus, Alaska: Implications for Determining the Effect of Human Caused Mortality on Population Size." MS thesis, Evergreen State College, Olympia, WA, 2013.

Prevéy, J. S., C. Rixen, N. Rüger, T. T. Høye, A. D. Bjorkman, I. H. Myers-Smith, S. C. Elmendorf, I. W. Ashton, N. Cannone, C. L. Chisholm et al. "Warming Shortens Flowering Seasons of Tundra Plant Communities." *Nature Ecology and Evolution* 3 (2019): 45–52.

Proctor, M. F., B. N. McLellan, G. B. Stenhouse, G. Mowat, C. T. Lamb, and M. S. Boyce. "Effects of Roads and Motorized Access on Grizzly Bear Populations in British Columbia and Alberta, Canada." *Ursus* (2019). https://doi.org/10.2192/URSUS-D-18-00016.2.

Pyke, G. H., H. R. Pulliam, and E. L. Charnov. "Optimal Foraging: A Selective Review of Theory and Tests." *Quarterly Review of Biology* 52 (1977): 137–154.

Ramey, A. M., C. A. Cleveland, G. V. Hilderbrand, K. Joly, D. D. Gustine, B. A. Mangipane, W. Leacock, A. Crupi, D. E. Hill, J. P. Dubey, and M. Yabsley. "Exposure of Alaska Brown Bears (*Ursus arctos*) to Bacterial, Viral, and Parasitic Agents Varies Spatiotemporally and May Be Influenced by Age." *Journal of Wildlife Diseases* 55, no. 3 (2019): 576–588. https://doi.org/10.7589/2018-07-173.

RAWS data. 2011–2021. Accessed August 1, 2022. https://wrcc.dri.edu/cgi-bin/rawMAIN.pl?akAFPK.

Ringsmuth, K. 2013. *At the Heart of Katmai: An Administrative History of the Brooks River Area.* Washington, DC: United States Department of the Interior, 2013.

Robbins, C. T., J. K. Fortin, K. D. Rode, S. D. Farley, L. A. Shipley, and L. A. Felicetti. "Optimizing Protein Intake as a Foraging Strategy to Maximize Mass Gain in an Omnivore." *Oikos* 116 (2007): 1675–1682. https://doi.org/10.1111/j.0030-1299.2007.16140.x.

Rode, K. D., S. D. Farley, and C. T. Robbins. "Behavioral Responses of Brown Bears Mediate Nutritional Effects of Experimentally Introduced Tourism." *Biological Conservation* 133 (2006): 70–80.

Rode, K. D., C. T. Robbins, and S. D. Farley. "Sexual Dimorphism, Reproductive Strategy, and Human Activities Determine Resource Use by Brown Bears." *Ecology* 87 (2006): 2636–2646.

Rogers, M. C. "Applications of Stable Isotope Analysis to Advancing the Understanding of Brown Bear Dietary Ecology." PhD diss., University of Alaska, Fairbanks, 2021.

Rogers, M. C., G. V. Hilderbrand, D. D. Gustine, K. Joly, W. B. Leacock, B. A. Mangipane, and J. M. Welker. "Splitting Hairs: Dietary Niche Breadth Modeling Using Stable Isotope Analysis of a Sequentially Grown Tissue." *Isotopes in Environmental and Health Studies* 56, no. 4 (2020): 358–369. https://doi.org/10.1080/10256016.2020.1787404.

Rogers, S. A., C. T. Robbins, P. D. Mathewson, A. M. Carnahan, F. T. van Manen, M. A. Haroldson, W. P. Porter, T. R. Rogers, T. Soule, and R. A. Long. "Thermal Constraints on Energy Balance, Behaviour, and Spatial Distribution of Grizzly Bears." *Functional Ecology* 35 (2021): 398–410.

Saltonstall, P., and A. Steffian. "Archaeological Survey in Eastern Alitak Bay." Report prepared for Kadiak, LLC, and the United States Fish and Wildlife Service, Alutiiq Museum and Archaeological Repository, Kodiak, AK, 2019.

Saxton, M. W. "Investigating Population Genetics and Seasonal Variation of Transcription in Brown Bears (*Ursus arctos*)." PhD diss., Washington State University, Pullman, 2021.

Schmidt, J. H., K. L. Rattenbury, H. L. Robison, T. S. Gorn, and B. S. Shults. "Using Non-Invasive Mark-Resight and Sign Occupancy Surveys to Monitor Low-Density Brown Bear Populations across Large Landscapes." *Biological Conservation* 207 (2017): 47–54.

Schmidt, J. H., H. L. Robison, L. S. Parrett, T. S. Gorn, and B. S. Shults. "Brown Bear Density and Estimated Harvest Rates in Northwestern Alaska." *Journal of Wildlife Management* 85, no. 2 (2021): 202–214.

Schmidt, J. H., T. L. Wilson, W. L. Thompson, and B. A. Mangipane. "Integrating Distance Sampling Survey Data with Population Indices to Separate Trends in Abundance and Temporary Immigration." *Journal of Wildlife Management* 86 (2022): e22185.

Schmidt, J. H., T. L. Wilson, W. L. Thompson, and J. H. Reynolds. "Improving Inference for Aerial Surveys of Bears: The Importance of Assumptions and the Cost of Unnecessary Complexity." *Ecology and Evolution* 7 (2017): 4812–4821.

Sellers, R. A., S. Miller, T. Smith, and R. Potts. *Population Dynamics of a Naturally Regulated Brown Bear Population on the Coast of Katmai National Park and Preserve.* Final report, NPS/AR-NRTR-99/36, 1999.

Servheen, C., S. Herrero, and B. Peyton. *Bears—Status Survey and Conservation Action Plan.* Gland, Switzerland: International Union for Conservation of Nature (IUCN) Bear Specialist Group, 1999.

Sheldon, C. *The Wilderness of Denali: Explorations of a Hunter-Naturalist in Northern Alaska.* New York: Derrydale, 1930.

Shepherd, T., and R. Frith. *Monitoring Visitor Use in the Southwest Alaska Network Using Commercial Use Authorization (CUA) Reports: Protocol Narrative Version 1.0.* Natural Resource Report NPS/SWAN/NRR—2018/1693. Fort Collins, CO: National Park Service, 2018.

Simon, J. J. "Customary and Traditional Use Worksheet, Black Bears, Game Management Units 12, 19, 20, 21 and 24 (Interior Alaska)." Prepared for the November 2008 Juneau Board of Game Meeting. Special Publication BOG 2008-07. Alaska Department of Fish and Game, Division of Subsistence, Juneau, 2008.

Singer, F. J., and J. Beattie. "The Controlled Traffic System and Associated Wildlife Responses in Denali National Park." *Arctic* 39 (1986): 195–203.

Skora, L. C. "Population Dynamics of Brown Bears along Brooks River in Katmai National Park, Alaska." MS thesis, Michigan State University, Lansing, 2021.

Smith, T. S. "Effects of Human Activity on Brown Bear Use of the Kulik River, Alaska." *Ursus* 13 (2002): 257–267.

Smith, T. S., S. Herrero, and T. D. DeBruyn. "Alaskan Brown Bears: Habituation and Humans." *Ursus* 16 (2005): 1–10.

Smith, T. S., S. Herrero, T. D. Debruyn, and J. M. Wilder. "Efficacy of Bear Deterrent Spray in Alaska." *Journal of Wildlife Management* 72 (2008): 640–645.

Smith, T. S., and B. A. Johnson. "Modeling the Effects of Human Activity on Katmai Brown Bears (*Ursus arctos*) through the Use of Survival Analysis." *Arctic* 57 (2004): 160–165.

Smith, T. S., and S. T. Partridge. "Dynamics of Intertidal Foraging by Coastal Brown Bears in Southwestern Alaska." *Journal of Wildlife Management* 68 (2004): 233–240.

Smith, T. S., J. M. Wilder, G. York, M. E. Obbard, and B. W. Billings. "An Investigation of Factors Influencing Bear Spray Performance." *Journal of Wildlife Management* 85 (2021): 17–26.

Sorum, M. S., M. D. Cameron, A. Crupi, G. K. Sage, S. L. Talbot, G. V. Hilderbrand, and K. Joly. "Pronounced Brown Bear Aggregation along Anadromous Streams in Interior Alaska." *Wildlife Biology* (2023): e01057. https://doi.org/10.1002/wlb3.01057.

Sorum, M. S., K. Joly, and M. D. Cameron. "Use of Salmon (*Oncorhynchus* spp.) by Brown Bears (*Ursus arctos*) in an Arctic, Interior, Montane Environment." *Canadian Field-Naturalist* 133, no. 2 (2019): 151–155. https://doi.org/10.22621/cfn.v133ix.2114.

Sorum, M. S., K. Joly, M. D. Cameron, D. D. Gustine, and G. V. Hilderbrand. "Salmon Sleuths: GPS-Collared Bears Lead Researchers to Unknown Salmon Streams in Interior Alaska." *Alaska Park Science* 19, no. 1 (2020): 4–9. https://www.nps.gov/articles/aps-19-1-2.htm.

Sorum, M. S., K. Joly, A. G. Wells, M. D. Cameron, G. V. Hilderbrand, and D. D. Gustine. "Den-Site Characteristics and Selection by Brown Bears (*Ursus arctos*) in the Central Brooks Range of Alaska." *Ecosphere* 10, no. 8 (2019): e02822. https://doi.org/10.1002/ecs2.2822.

Støen, O.-G., E. Bellemain, S. Sæbø, and J. E. Swenson. "Kin-Related Spatial Structure in Brown Bears, *Ursus arctos*." *Behavioral Ecology and Sociobiology* 59 (2005): 191–197.

Stokes, J. *Natural Resource Utilization of Four Upper Kuskokwim Communities*. Technical Paper 86. Juneau: Alaska Department of Fish and Game, Division of Subsistence, 1985.

Swanson, D. K. *Vegetation and Snow Phenology Monitoring in the Arctic Network through 2020: Results from Satellites and Land-Based Cameras*. Natural Resource Report. Anchorage: National Park Service, 2021.

Swanton, J. R. "Social Condition, Beliefs, and Linguistic Relationships of Tlingit Indians." In *26th Annual Report, Bureau of American Ethnology*, 391–485. Washington, DC: U.S. Government Printing Office, 1908.

Swanton, J. R. *Tlingit Myths and Texts*. Smithsonian Institution, Bureau of American Ethnology, Bulletin 39. Washington, DC: U.S. Government Printing Office, 1909.

Swenson, J. E., F. Sandegren, S. Brunberg, and P. Wabakken. "Winter Den Abandonment by Brown Bears *Ursus Arctos*: Causes and Consequences." *Wildlife Biology* 3 (1997): 35–38.

Sytsma, M. T., T. Lewis, B. Gardner, and L. R. Prugh. "Low Levels of Outdoor Recreation Alter Wildlife Behavior." *People and Nature* 4 (2022): 1547–1559. https://doi.org/10.1002/pan3.10402.

Tape, K. D., D. D. Gustine, R. W. Ruess, L. G. Adams, and J. A. Clark. "Range Expansion of Moose in Arctic Alaska Linked to Warming and Increased Shrub Habitat." *PLoS ONE* 11 (2016): 1–12.

Thornton, T. F. *Subsistence Use of Brown Bear in Southeast Alaska*. Technical Paper 214. Juneau: Alaska Department of Fish and Game, Division of Subsistence, 1992.

Troyer, W. *Distribution and Densities of Brown Bear on Various Streams in Katmai National Monument*. Anchorage: National Park Service, 1980.

Troyer, W. *Into Brown Bear Country*. Fairbanks: University of Alaska Press, 2005.

Troyer, W. *Movements and Dispersal of Brown Bear at Brooks River, Alaska*. Anchorage: National Park Service, 1980.

Troyer, W., and J. B. Faro. *Aerial Survey of Brown Bear Denning in the Katmai Area of Alaska*. Anchorage: National Park Service, 1975.

Trujillo, S. M., E. A. McKenney, G. V. Hilderbrand, L. S. Mangipane, M. C. Rogers, K. Joly, D. D. Gustine, J. A. Erlenbach, B. A. Mangipane, and D. J. R. Lafferty. "Correlating Gut Microbial Membership to Brown Bear Health Metrics." *Scientific Reports* 12 (2022): 15415. https://doi.org/10.1038/s41598-022-19527-4.

Trujillo, S. M., E. A. McKenney, G. V. Hilderbrand, L. S. Mangipane, M. C. Rogers, K. Joly, D. D. Gustine, J. A. Erlenbach, B. A. Mangipane, and D. J. R. Lafferty. "Intrinsic and Extrinsic Factors Influence on an Omnivore's Gut Microbiome." *PLoS ONE* 17, no. 4 (2022): e0266698. https://doi.org/10.1371/journal.pone.0266698.

Turner, C. "Determining the Effectiveness of Park Management Strategies at a Coastal Brown Bear Viewing Site in Katmai National Park, Alaska." Master's thesis, Royal Roads University, Victoria, BC, 2012.

United States Congress. The Alaska National Interest Lands Conservation Act, Public Law 96–487. December 2, 1980.

United States Congress. The National Park Service Organic Act, 16 U.S.C. 1916.

United States Department of the Interior, National Park Service. *Management Policies 2006*. Washington, DC: U.S. Government Printing Office, 2006.

United States Department of the Interior, National Park Service. *Final Environmental Impact Statement: Brooks River Area Development Concept Plan*. Juneau: Katmai National Park and Preserve, AK, 1996.

Van Daele, L. J., J. R. Morgart, M. T. Hinkes, S. D. Kovach, J. W. Denton, and R. H. Kaycon. "Grizzlies, Eskimos, and Biologists: Cross-Cultural Bear Management in Southwest Alaska." *Ursus* 12 (2001): 141–152.

Van de Kerk, M., S. Arthur, M. Bertram, B. Borg, J. Herriges, J. Lawler, B. Mangipane, C. Lambert Koizumi, B. Wendling, and L. Prugh. "Environmental Influences on Dall's Sheep Survival." *Journal of Wildlife Management* 84 (2020): 1127–1138.

Van Valen, L. "Morphological Variation and Width of Ecological Niche." *American Naturalist* 99 (1965): 377–390.

von Biela, V., L. Bowen, S. D. McCormick, M. P. Carey, D. S. Donnelly, S. Waters, A. M. Regish, S. M. Laske, R. J. Brown, S. Larson et al. "Evidence of Prevalent Heat Stress in Yukon River Chinook Salmon." *Canadian Journal of Fisheries and Aquatic Sciences* 77, no. 12 (2020): 1878–1892. https://doi.org/10.1139/cjfas-2020-0209.

Welch, C. A., J. Keay, K. C. Kendall, and C. T. Robbins. "Constraints on Frugivory by Bears." *Ecology* 8 (1997): 1105–1119.

White, J. "The Way of the Hunter: An Interview with Richard Nelson." *Sun Magazine* (May 1992).

White, L. "Huna Tlingit Place Names Recordings." Tapes 3 and 4. January 3, 2009. Glacier Bay National Park and Preserve Collection, Gustavus, AK.

White, L. Interview with M. B. Moss. Hoonah Indian Association, Chookaneidí, Hoonah, AK. September 21, 2019.

White, L., and A. Marvin. "Huna Clans and Marriage, Tape 1." March 9, 1993. Huna Heritage Foundation Collection, Juneau.

White, P. J., K. A. Gunther, and F. T. van Manen. *Yellowstone Grizzly Bears: Ecology and Conservation of an Icon of Wilderness.* Bozeman, MT: Yellowstone National Park, Yellowstone Forever, and U.S. Geological Survey, Northern Rocky Mountain Science Center, 2017.

Wilker, G. A., and V. G. Barnes Jr. "Responses of Brown Bears to Human Activities at O'Malley River, Kodiak Island, Alaska." *Ursus* 10 (1998): 557–561.

Woodhouse-Beyer, K. "Artels and Identities: Gender, Power, and Russian America." In T. L. Sweely, ed., *Manifesting Power: Gender and the Interpretation of Power in Archaeology*, 129–154. New York: Routledge, 2012.

Young, K. B., and T. M. Lewis. *Bears of Glacier Bay National Park and Preserve: A Summary of Bear Management in 2020 and 2021.* Resource Brief. Gustavus, AK: National Park Service, 2021.

INDEX

ABOUT THE AUTHORS

John Branson grew up in Maine and developed an affinity for the outdoors at an early age. A trip to Southeast Alaska in his youth exposed him to vast, wild places and planted the seed for what would become a life in Alaska beginning in 1969. During his time in Alaska, John has worked as a teacher, guide, caretaker, maintenance worker, and ranger, but found his calling as an historian for Lake Clark National Park and Preserve in 1989. In that role, he has tirelessly worked to document the history of Lake Clark and Bristol Bay, resulting in the publication of 12 books to date. John lives in Port Alsworth, Alaska, where he continues to explore, learn, and share his passion for Bristol Bay.

Matthew (Matt) D. Cameron is a wildlife biologist with the National Park Service, working primarily in Gates of the Arctic Park and Preserve and Yukon-Charley Rivers National Preserve. Since 2015, he has worked on research and management of caribou, brown bears, moose, and gray wolves with the National Park Service. He earned his PhD in wildlife biology from the University of Alaska Fairbanks.

Nina Chambers is a science communicator for the National Park Service in Alaska. The first part of her career was focused on landscape-scale conservation, and she transitioned in 2011 to an emphasis on science communication. She is managing editor of *Alaska Park Science*, a peer-reviewed science journal

◄ A sow bear and her cubs in the boreal forest. (NPS/Erika Jostad)

of the National Park Service Alaska Region. Through this and other media, she shares research findings with the public to inform decisions and increase public awareness of and engagement in science. Her first degree was in wildlife biology, and she earned her MS in wildland management from the University of Idaho.

Amy Craver has worked as a cultural anthropologist and subsistence coordinator for Denali National Park and Preserve since 2002. Prior to working with Denali, she worked with the Office of Subsistence Management. She has conducted extensive ethnographic and oral history fieldwork in northwest Alaska and in the Aleutians. She grew up in Talkeetna and earned an MA in folklore from Indiana University.

William (Will) Deacy is a landscape ecologist at Rocky Mountain National Park, working on research and management of elk and moose within the Ecosystem Balance Program. From 2019 through 2023, he was the brown bear and Dall's sheep vital sign lead for the Arctic Inventory and Monitoring Network based in Fairbanks, Alaska. From 2012 through 2018, Will conducted research on brown bear and salmon ecology in Kodiak, Alaska. During this work, he received his PhD in systems ecology from the University of Montana.

Dael Devenport has been an archaeologist in Alaska since 1999. She has worked for Native corporations, tribes, and the federal government. She currently works for the National Park Service in Anchorage. She earned an interdisciplinary anthropology MS from the University of Alaska Fairbanks Resilience and Adaptation Program.

Joy Erlenbach is a wildlife biologist for the United States Fish and Wildlife Service at Kodiak National Wildlife Refuge. She completed her PhD work studying bears in coastal Katmai National Park and has been a student of bear behavior and diet since 2007.

Karen Evanoff is Dena'ina Athabascan from Nondalton. Karen has a degree in anthropology and works for Lake Clark National Park and Preserve as the cultural anthropologist. The majority of her work involves working with Alaskan Native people on projects or programs relating to cultural preservation and revitalization. She is studying a new method of research that involves an integral or holistic approach, thereby working from a broader framework for research.

Victoria Florey is a subsistence program analyst for the National Park Service in Alaska. Victoria was born in Ninilchik, Alaska, where she spent her summers fishing and autumns moose hunting and berry picking. Her focus is on providing support to Alaska parks in meeting their subsistence-related programmatic needs. She also coordinates the Alaska Native Science and Engineering Program internship program for the National Park Service Alaska Region. She earned her BS in anthropology from the University of Alaska Fairbanks and her MPhil in archaeology from Cambridge University, England.

Susan Georgette worked for the United States Fish and Wildlife Service, Selawik National Wildlife Refuge for 15 years, retiring as refuge manager in 2022. Prior to that, Susan conducted social science research for many years for the Division of Subsistence, Alaska Department of Fish and Game—recording Indigenous knowledge of resources, shaping regulations suited to traditional practices, and documenting subsistence economies. She has lived and worked in Kotzebue, Alaska, for more than 30 years. She holds a BA in environmental studies from the University of California Santa Cruz.

Kelsey Griffin is director of the Ocean Alaska Science and Learning Center and was previously a wildlife biologist for the National Park Service at Katmai National Park and Preserve in Alaska. She has worked on bear research since 2016 along with other species that use the coastal environment, including wolves and seabirds. She has a background in wildlife ecology and management, including human dimensions and conservation education. She earned her MS in wildlife, sustainability, and ecosystem sciences from Tarleton State University.

David (Dave) D. Gustine started as the bison conservation coordinator for the National Park Service in 2024. Since 1998, he has worked as a biologist in various capacities for state and federal agencies with a focus on the management of and research on mammals, including mule deer, wild horses, caribou, Stone's sheep, muskoxen, wolves, foxes, and bears—brown, black, and polar bears. Since 2010, he has worked for the United States Geological Survey in Alaska and the National Park Service in Grand Teton National Park, as well as led the Polar Bear Program for the United States Fish and Wildlife Service. Dave earned his PhD in biological sciences from the University of Alaska Fairbanks.

Troy Hamon was the chief of resource management and science for Katmai National Park and Preserve for more than a decade. Currently, he is the park's pilot in support of wildlife management and park stewardship. He earned his PhD from the University of Washington.

Grant V. Hilderbrand works for the National Park Service in the Alaska region and has studied bears, wolves, and salmon since 1993. He is a member of the International Union for the Conservation of Nature (IUCN) North American Bear Expert Team, was chair of the 2016 International Conference on Bear Research and Management, holds affiliate faculty appointments at several universities, and is an active member of the Wildlife Society. Grant earned his PhD in zoology from Washington State University.

Kim A. Jochum works for the National Park Service in the Alaska Regional Subsistence Program. Kim studied brown bears in the Russian Far East and Alaska and polar bears in Canada. She led a wildlife management program in interior Alaska managing human-bear encounters and researching Dall's sheep, bats, shorebirds, and salmon, among others. Kim is an officer of the Wildlife Society Alaska Chapter. She earned an interdisciplinary PhD in biological sciences at the University of Alaska Fairbanks, applying a social–ecological systems approach to human-bear encounters across the Pacific Rim.

Kyle Joly is a wildlife biologist for the National Park Service, working in Gates of the Arctic Park and Preserve, Yukon-Charley Rivers National Preserve, and other northern Alaska parks. He started working for the National Park Service in Alaska in 1994 and is primarily focused on the management, research, and conservation of large mammals, including caribou, brown bears, gray wolves, moose, and Dall's sheep. He earned his PhD in wildlife biology from the University of Alaska Fairbanks.

Diana Lafferty is an associate professor in the Department of Biology at Northern Michigan University in Marquette, where her research program focuses on advancing understanding of how wildlife populations and their associated communities and ecosystems respond to human-mediated global change. In this pursuit, she has studied a diverse array of mammals, including brown bear, black bear, fisher, marten, mink, wild boar, and snowshoe hare. Diana earned her PhD in forest resources from Mississippi State University.

Tania Lewis is the terrestrial wildlife biologist for Glacier Bay National Park and Preserve. She has been studying bears in Glacier Bay since 2001 and leading the park's Human-Bear Management Program since 2006. Tania studies multiple wildlife species to document the ecological effects of changing climate and landscape, determine animal colonization patterns in recently deglaciated areas, and mitigate negative human impacts on wildlife. She earned her MS in wildlife biology from the University of Alaska Fairbanks.

Buck Mangipane is a natural resource program manager for the National Park Service, working in Lake Clark National Park and Preserve. He began working for the National Park Service in 2002 and has taken part in the management of and research on caribou, Dall's sheep, brown bears, black bears, wolves, moose, and bald eagles. Buck earned his BS in wildlife biology from Colorado State University.

Lindsey Mangipane is a wildlife biologist with the United States Fish and Wildlife Service polar bear team, where her work primarily focuses on preventing and addressing human–polar bear conflicts. Before joining the USFWS, Lindsey worked with Montana Fish, Wildlife, and Parks, where she helped resolve conflicts between grizzly bears and people. Lindsey attended Mississippi State University, where she completed her master's thesis on evaluating the effects of dietary plasticity and landscape heterogeneity on brown bears in Lake Clark National Park and Preserve.

Rachel Mason is the senior cultural anthropologist for the Alaska region of the National Park Service. She has also worked for the Division of Subsistence, Alaska Department of Fish and Game, and the Office of Subsistence

Management, United States Fish and Wildlife Service. She has a PhD in anthropology from the University of Virginia.

Mary Beth Moss is the tribal liaison at Glacier Bay National Park and Preserve. She previously served as the chief of resource management at Glacier Bay and as the tribal administrator and cultural resource specialist for the Hoonah Indian Association (HIA), a federally recognized Tribal government. She was adopted by the Sik'nax̲.ádi and Chookaneidí Clans and was naturalized as a Tribal citizen by HIA in 2021. She received a BS in forest science from Penn State University and an MS in wildlife ecology from Virginia Tech.

Marcy Okada is the subsistence coordinator for Gates of the Arctic National Park and Preserve and the Yukon-Charley Rivers National Preserve. She has worked for the National Park Service since 2009 and oversees the subsistence management program, which includes government-to-government consultation and outreach to affiliated local rural communities. Marcy earned her MS in natural resources management from the University of Alaska Fairbanks.

Jack Omelak works as a cultural anthropologist for the National Park Service in the Alaska Regional Office. Jack serves as the regional Native American Graves Protection and Repatriation Act coordinator and as the regional ethnographic inventory database coordinator. Previously, he worked extensively with and for Alaska Native organizations.

Patricia (Pat) Owen is a wildlife biologist for the National Park Service, working in Denali National Park and Preserve. She started working for Denali in 1988 and oversees the park's wildlife management program to minimize human-wildlife conflicts by providing wildlife safety information to visitors and actively managing wildlife, when necessary. She focuses primarily on the management, research, and conservation of grizzly bears and other large mammals. Pat earned her MS in biology from East Stroudsburg University in Pennsylvania.

Dillon Patterson works in the subsistence program at the National Park Service Alaska Regional Office. He started working for the NPS in 2020 as a Pathways Program anthropology student. In this role, Dillon is also pursuing a PhD in environmental anthropology from the University of Connecticut. His dissertation research focuses on traditional ecological knowledge on caribou in Katmai National Preserve and the role of traditional knowledge in federal subsistence management.

David Payer was the regional wildlife biologist for the National Park Service in Alaska. Starting in 1999, David conducted and coordinated research on Arctic and sub-Arctic wildlife, with a focus on effects of climate change and anthropogenic stressors. David earned a PhD in wildlife ecology from the University of Maine and a doctorate in veterinary medicine from Cornell University.

Michael Saxton is a wildlife biologist at Katmai National Park and Preserve in Alaska. He has worked with federal and state agencies, as well as universities and private consulting firms, promoting the conservation of wildlife. Michael began working with bears in 2008 and has focused on Alaskan coastal brown bear research and management since joining the Katmai team in 2012. Michael earned his PhD in biological sciences from Washington State University.

Joshua (Josh) H. Schmidt is a biometrician for the National Park Service's Central Alaska Network. Since 2008, he has worked to develop efficient and effective survey designs for a variety of species and systems, including an emphasis on landscape-scale surveys for large mammals. A portion of this work has focused on the development of survey techniques applicable to brown bear populations throughout Alaska's parks. Josh earned his PhD in biological sciences from the University of Alaska Fairbanks.

Andee Sears has worked for the National Park Service for more than 2 decades. She started as a summer backcountry ranger in Olympic National Park in Washington State. In the years that followed, Andee worked in Wrangell–St. Elias National Park and Preserve, Denali National Park and Preserve, and Yosemite National Park, among other places, as a park ranger and a criminal investigator. She received her law degree and certificate in environmental law from the University of Utah, where she also served as editor-in-chief of the *Journal of Land,*

Resources, and Environmental Law. Her primary areas of focus on policy matters have been wildlife management, authority and jurisdiction over waters, and motorized access in wilderness. Andee serves as the regional chief ranger for the National Park Service Alaska region. In this capacity, she provides advice on a wide range of resource protection matters, including wildlife management.

Leslie Skora has been a wildlife biologist at Katmai National Park and Preserve since 2013. Her work focuses on monitoring wildlife abundance and population dynamics within the park. Leslie earned her MS in fisheries and wildlife at Michigan State University and is a PhD student in environmental conservation at the University of Massachusetts Amherst.

Mathew (Mat) S. Sorum is a wildlife biologist for the National Park Service in northern Alaska and has studied brown bears and wolves since 2008. Mat leads the wolf monitoring program in Yukon-Charley Rivers National Preserve. He earned his graduate degree from the University of Idaho while studying brown bear movement and diet patterns on the Kodiak National Wildlife Refuge.

Pam Sousanes is a physical scientist for the National Park Service, working in Alaska's 8 northernmost national parks. She started working for the National Park Service in 1993. She manages a climate monitoring program that covers 40 million acres of high-latitude parklands, with a goal of documenting the weather and climate trends that are impacting park landscapes and ecosystems. Pam earned her undergraduate degree in environmental conservation from the University of Colorado, Boulder, and a master's degree in ecological management from the Harvard Extension School.

Laura Stelson is a PhD student and instructor in the programs for anthropology and human dimensions of natural resources and the environment at Penn State University. Laura has spent much of her time working on archaeological projects in Germany, Central America, and throughout the western United States. She started working in Katmai National Park in 2016. Her research interests include the application of remote sensing and geochemical material sourcing to the reconstruction of past environments and the ways people interacted with them. She received her MA in anthropology from the University of Bonn (Germany).

David K. Swanson is a retired ecologist for the National Park Service's Arctic Network, which includes the 5 parks in northern Alaska. He was responsible for monitoring change in landscape features such as vegetation, phenology (the seasonal timing of the snow and growing seasons), permafrost, and landforms. He has an MS in geology from the University of Colorado and a PhD in soil science from the University of Minnesota. David previously worked as a soil scientist and ecologist for the United States Department of Agriculture–Natural Resources Conservation Service and the United States Forest Service.